British Romanticism, Climate Change, and the Anthropocene

David Higgins

British Romanticism, Climate Change, and the Anthropocene

Writing Tambora

palgrave
macmillan

David Higgins
School of English
University of Leeds
Leeds, UK

ISBN 978-3-319-67893-1 ISBN 978-3-319-67894-8 (eBook)
DOI 10.1007/978-3-319-67894-8

Library of Congress Control Number: 2017953777

Printed on acid-free paper

This Palgrave Macmillan imprint is published by Springer Nature
The registered company is Springer International Publishing AG
The registered company address is: Gewerbestrasse 11, 6330 Cham, Switzerland

To Edward, with love and hope

Acknowledgements

This book was started during a period of study leave funded by the School of English at the University of Leeds and completed while I held an Arts and Humanities Research Council Leadership Fellowship. I am very grateful to both institutions for funding this research and for the support of my fellow Romanticist and former Head of School, John Whale. I have benefited from the excellent research environment at Leeds, where I have worked with brilliant colleagues in the environmental humanities and Romantic studies. I must particularly thank Amy Cutler, David Fairer, Richard De Ritter, Carl McKeating, and Sebastien Nobert. I could not have found a better mentor for developing projects in the environmental humanities than Graham Huggan. Jeremy Davies has taught me most of what I know about the Anthropocene; he also read a complete draft of the book and commented on it with his usual rigour, sympathy, and perceptiveness. I have learnt much from the MA students on our Romantic Ecologies module, especially Izzy Gahan. I owe a great deal to the terrific research assistance of Tess Somervell. She provided significant support with the intellectual and technical aspects of the book, and made the writing process much quicker and easier than it would otherwise have been. I have also received help from many colleagues at other institutions, including Eric Gidal, Evan Gottlieb, Dehlia Hannah, Ian Haywood, Adeline Johns-Putra, Tobias Menely, Anna Mercer, Susan Oliver, Kate Rigby, and Jane Stabler. It is always a pleasure to work with my editor at Palgrave, Ben Doyle, and the rest of his editorial team: Milly Davies, Eva Hodgkin, and Tomas Rene. The book's two anonymous reviewers offered valuable encouragement and critique. The late Anthony Carrigan motivated and inspired me during the early stages of the project and I wish that he was still around to see how much it owes to his superb example.

Writing Tambora has taken me to some fascinating places, including the darker reaches of nihilistic and antinatalist thought. It's proved difficult to reconcile the ideas that I have encountered with the profound joy of becoming a parent during the same period and sharing this experience with my always supportive and insightful partner Alys. This book is dedicated to our son, Edward James Higgins.

Contents

CHAPTER 1

Introduction

Abstract The introduction positions the book in relation to extant scholarship, particularly on the Tambora eruption, material ecocriticism, and the Anthropocene. *Writing Tambora* innovatively addresses the question of how global catastrophe is rhetorically produced. The subsequent discussion details three areas of research which the book draws on and to which it also contributes: disaster studies; new materialism and speculative realism; and the cultural history of climate change. Challenging the idea of the Anthropocene as an epistemological breach between past and present, *Writing Tambora* instead articulates a more historicist methodology which finds important resonances between Romantic and present-day imaginings of climate change—especially around questions of agency—and therefore brings Romantic studies and the environmental humanities into productive dialogue.

Keywords Material ecocriticism · Speculative realism · New materialism · Disaster studies · Anthropocene · Climate change

On the evening of 5 April 1815, the inhabitants of Java heard a number of explosions that continued intermittently until the following day. At first they were 'almost universally attributed to distant cannon,' but in fact this was the opening salvo in the eruption of Mount Tambora on the isle of Sumbawa, hundreds of miles to the east.[1] A hazy atmosphere and slight fall of ash followed over several days. At about 7 pm on 10 April, the

D. Higgins, *British Romanticism, Climate Change, and the Anthropocene*,
DOI 10.1007/978-3-319-67894-8_1

mountain blew up. According to the sole eyewitness account, 'three distinct columns of flame burst forth near the top [...] In a short time the whole Mountain [...] appeared like a body of liquid fire extending itself in every direction.'[2] The explosions could be heard over 2000 kilometres from the eruption. Due to the huge amounts of volcanic material emitted, 'many places within a 600-kilometre radius remained pitch black for a day or two' and the ash fall affected a much larger area.[3] This was one of the very largest documented eruptions of the Holocene period—far bigger than the much better known Krakatoa eruption of 1883—and it had devastating consequences for local populations.[4] It wiped out the kingdom of Tambora, and the ash destroyed agriculture and contaminated drinking water across Sumbawa and nearby islands. The exact death toll from the explosions, pyroclastic currents, tsunami, and local famine and disease is impossible to know, but plausible estimates put it at between 60,000 and 120,000 people across Sumbawa, Bali, and possibly other parts of the archipelago such as Lombok and eastern Java.[5] The huge amount of sulphur released into the atmosphere formed a sulphuric acid aerosol that affected world climate patterns, leading to a global cooling of between one and two degrees centigrade (strongest in the northern hemisphere) and unusual weather in the period from 1816 to 1818.[6] In particular, 1816 became known in Europe and North America as the 'Year without a Summer' due to unseasonably cold and wet weather.[7] Following the groundbreaking work of John D. Post, scholars have identified Tambora as a key factor in the harvest failures and food scarcities across the globe in the late 1810s, and perhaps even the typhus and cholera epidemics of the period.[8] The history of the eruption shows on a global scale the catastrophic consequences of a powerful natural hazard in combination with large numbers of people made vulnerable by their poverty.

It would not be until the twentieth century that meteorologists made the connection between Tambora and the weather conditions of the late 1810s, and at the time the eruption was not widely reported. However, in the last three decades, the story of Tambora and its effects, local and global, has been told several times.[9] The present study takes a new approach by addressing the eruption and the subsequent global climate crisis as a textual catastrophe. To do so is not to downplay the intense human suffering to which it contributed, or its profound environmental effects. Rather, I draw attention to how Tambora, and other such catastrophes, are productively understood as processes in which the material and the discursive are intertwined. I also reveal how some canonical Romantic works written in the period from 1815 to 1818 are more generally concerned with climate change. Despite obvious differences between Romantic ideas and contemporary environmental issues, this book points to some important connections between them. If Romantic writers were often excited by the human capacity to shape the world, they were also troubled by our vulnerability to elemental forces apparently beyond our control. They therefore speak to the increasingly influential idea that we are entering a new geological epoch characterised by the complex entanglement of human activity and earth systems: the Anthropocene.[10] The term is not without controversy. Some scholars have criticised it for suggesting a species-wide agency that obscures the inequalities that have driven, and are driven by, global environmental change.[11] Others have suggested that it implies a dangerous anthropocentrism; Stacy Alaimo, for example, suggests that its 'arrogance [...] need be held in check by an elemental sense of the world as also, simultaneously, that which cannot be accessed, understood, and fundamentally altered by human practices.'[12] From an entirely different perspective, its use by non-scientists has been criticised as unhelpfully vague and detached from the specifics of the term as a stratigraphic marker that indicates a step change in Earth systems.[13] The start date has also been debated, with suggestions going as far back as the climatic impacts of agriculture ten thousand years ago and as far forward as the traces in the rock record left by nuclear testing after World War II.[14] Indeed, some leading geoscientists have identified the emergence of the Anthropocene in the Romantic period, noting the stratigraphic potential of the coincidence of the Tambora eruption and the early stages of industrial capitalism.[15] This book examines three case studies: the narrative of the eruption and its aftermath compiled by the British administration in Java; the works of Byron and the Shelleys in the summer of 1816; and political writing in Romantic periodicals about the 'distresses' of 1816–1817. Through these case studies, this book shows how British Romantic texts speak to topics of pressing significance in the Anthropocene: the relationship between human and nonhuman agency; the precarity of human life on an increasingly volatile planet; and the interplay of individual consciousness, political structures, and earth systems.

Writing in 2011, Timothy Clark pointed to the 'relative absence in ecocriticism' of the topic of climate change.[16] It would be difficult to justify such a claim now. In recent years, there has been a substantial development of scholarship in the field, particularly around climate change fiction.[17] There has also been increasing interest among researchers in how climate change has been experienced and represented in earlier periods.[18] The humanities are belatedly beginning to wake up to the significance and urgency of climate change as a topic, and the need for disciplines to work together in order to try to combat the short-termism and denialism that pervades much political and corporate discourse. This awareness is reflected not only within specialist academic writing, but in more publicly focused works, including two recent books on Tambora and the 'Year without a Summer.'[19] Given these publications, and the earlier identification of the relationship between Tambora and Romantic literature by scholars such as Jonathan Bate and Alan Bewell, it might well be asked if much more remains worth saying on the subject.[20] *Writing Tambora* takes a new approach through close analysis of a wide range of sources within an intellectual framework provided by the emergence of the environmental humanities. I read the writings 'produced' by Tambora not so much as illustrative of its effects, but as themselves part of the intersection of culture and matter that creates an environmental catastrophe. My approach builds on recent work in 'material ecocriticism,' which emphasises the 'constitutive engagement of human discursive systems with the material world.'[21] The composition of eyewitness accounts of the eruption and its effects put together under the auspices of Sir Stamford Raffles tells us much about Tambora's physical processes and has therefore been of great value to vulcanologists.[22] Similarly, the 1816 writings of the Diodati Circle tell us much about the weather conditions of 1816 and their social effects. However, more noteworthy from a humanities perspective is the complexity with which these sources mediate human-material entanglements: a complexity that is best apprehended through attentiveness to textual detail and ambivalence.

This attentiveness distinguishes the present study from earlier scholarship on Tambora. Even work on the topic by literary critics, including Gillen D'Arcy Wood's excellent study, has tended to take a cultural-historical approach that uses quotations for illustrative rather than analytical purposes. Therefore, the full implications of the Tambora crisis for how we understand the relationship between humans and nonhuman nature have not been addressed. This book does not seek to tell the story of Tambora, but to address the question of how global catastrophe is rhetorically produced. Its key case studies are collaborative and heteroglossic, suggesting that the writing of a catastrophe of such scale and power goes beyond what can be achieved by any single author. My methodology is informed by a belief that the best way for literary scholars to contribute to the environmental humanities is not by trying to be sociologists or cultural historians, but by harnessing our focus on textuality within an interdisciplinary framework. The remainder of this introduction addresses three interdisciplinary contexts for the discussion that follows: humanities approaches to disaster; the new materialism; and the cultural history of climate change. By bringing these contexts to bear on its case studies, this book not only sheds new light on Romantic constructions of ecological catastrophe, but also brings Romantic studies and the environmental humanities into dialogue.

Environmental Catastrophe and Disaster

Maurice Blanchot's *The Writing of the Disaster* (1980) meditates on the superfluity and the necessity of writing in the context of the human-made disasters of the twentieth century. Although not explicitly concerned with ecology, it also offers ways of thinking about so-called 'natural' disasters.[23] According to Blanchot, 'the disaster de-scribes,' pushing writing to its 'limit' by going beyond what our everyday epistemologies can contain.[24] Its scale and traumatic effects threaten to make words meaningless, to 'lay waste' to them.[25] And yet disaster is also 'the force of writing' and does not lie beyond it.[26] Disaster is a kind of present absence that requires writing even as it undermines and 'eludes' it.[27] And writing is itself a kind of disaster, a violent eruption that crosses the boundary between being and world.[28] This idea of writing is echoed in Deleuze and Guattari's claim that 'the book is not an image of the world. It forms a rhizome with the world'.[29] That is, texts do not *represent* reality but manifest a fluid set of connections with it. Such theorising may seem a long way from the technocratic field of disaster studies. But it has long been understood that what we call a 'natural disaster' or an 'environmental catastrophe' actually springs from 'the intersection of natural hazards such as earthquakes, hurricanes, and volcanic eruptions with human populations in varying states of economic, social, and cultural vulnerability.'[30] Therefore, as Mark Anderson suggests, disasters that are often seen as external insurgencies into the human world in fact 'embody yet another facet of human interaction with the environment and as such must be mediated through culture.'[31] Such mediations are increasingly of interest to scholars in the environmental humanities and there has been valuable work on recent disasters such as the Indian Ocean earthquake and tsunami of 2004 and Hurricane Katrina.[32] Much of this scholarship is concerned with what Anderson calls 'the cultural politics of catastrophe': the ways in which 'disaster narratives serve to legitimize and to delegitimize political discourse,' and the present study draws on this approach.[33] Thus in Chap. 2, I show how the report on the Tambora eruption compiled by the British imperial administration in Java is inextricable from the brief history of that administration and the wider context of Britain's global conflict with France. Chapter 3 addresses how the Diodati Circle's writings of 1816 challenge conservative and providential understandings of the intertwining of natural and human history, while also showing the limitations of emancipatory political projects that do not recognise the finitude of the human species. For Byron and the Shelleys, imagined catastrophes threaten to 'delegitimize' *all* political discourse by revealing humanity's transience, vulnerability, and contingency. And my final chapter shows how the unusual weather of 1816–1817 was evoked in British periodicals in order to legitimise both conservative and reformist responses to a nation in crisis.

Humanities approaches to the cultural politics of catastrophe have much to offer the field of disaster studies which, as Anthony Carrigan has pointed out, has tended to be affiliated with 'normative politics' and technocracy.[34] He calls for 'a self-conscious reformulation of disaster studies methods,' which must include 'a more rigorous understanding of how narratives shape our perception and understanding of what constitutes a disaster.'[35] Some of the most exciting work in the environmental humanities is contributing to the development of such an understanding. Rob Nixon brings together 'reflections on empire, foreign policy, and resistance with questions about aesthetic strategy' in his influential study of disasters that are attritional and concealed rather than swift and spectacular.[36] Timothy Clark has shown how reading literary texts with attention to how they work contradictorily across different scales sheds light on 'the slow-motion catastrophes of international capitalism' and undermines the 'easy, daily equations of moral and political accounting' in the context of the Anthropocene.[37] And, like the present study, Kate Rigby's *Dancing with Disaster* understands disasters as 'material-discursive processes.'[38] Bringing together a range of texts from the Romantic period to the present, including Mary Shelley's *The Last Man* (1826), Rigby argues for the value of literature (and especially fiction) in 'hold[ing] up to scrutiny the kinds of often contradictory stories that people tell about collective calamities involving the dynamic interaction of diverse human and nonhuman agencies and processes.'[39] *Writing Tambora* draws on this valuable body of research while also offering something new through its intense focus on a crucial moment in the history of discursive-material interactions. This focus allows for a deep analysis of the relations between different sorts of texts that does not privilege narrative fiction; perhaps partly as a result of its interest in multiple genres, the book has more to say about human marginality than much work in the environmental humanities.

Generally speaking, this book uses the term 'catastrophe' rather than 'disaster.'[40] Whereas the latter word suggests an unfortunate and predestined occurrence, catastrophe—from the Greek καταστροφή (an overturning, a sudden turn, a conclusion)—indicates a major shift in the state of things that may well be destructive, but is not necessarily so. It is less anthropocentric than the word 'disaster' because it does not necessarily judge an event or process by whether its effects on human beings are positive or negative. As a term, it is equally applicable to the heat-death of the entire universe in Byron's 'Darkness' and the joyous apocalypse at the end of Percy Bysshe Shelley's *Prometheus Unbound* in which the Earth celebrates its freedom from Jupiter's tyranny. As Rigby points out, in Aristotle's *Poetics*, 'the *katastrophē* is intimately associated with *anagnorisis*: the moment of realization.' Catastrophes may offer 'opportunities for deeper understanding and, potentially, new directions.'[41] Furthermore, Timothy Morton has argued with his customary brio that the term '"ecological disaster" is an oxymoron.'[42] Given the etymology of the term—'an astrological misfortune' (*dis-astron*)—it only makes sense in the context of a worldview in which human activities take place against 'the star-studded dome of the world.' However, 'ecological science and ecological awareness force upon humans the collapse of any significant background or horizon.'[43] If ecology increasingly emphasises interconnectedness between the biosphere and the non-biosphere (what Morton elsewhere calls 'the mesh'), then the idea of a disaster as an incursion from the outside does not hold.[44] The term 'catastrophe' avoids this issue, although its association with 'suddenness' may exhibit one of Morton's problems with 'disaster'—'the tropology of an absolute end, a sudden misfortune'—which does not necessarily reflect the gradual and hidden effects of a process like global warming.[45] It is important that 'catastrophe' is used, therefore, with attentiveness to the different temporal scales through which human-environmental interactions take place. After all, the emergence of the Anthropocene has been relatively gradual in relation to human history and catastrophically fast in relation to the Earth system.

'Catastrophe' was an important trope in the Romantic period, for a number of reasons beyond the Tambora eruption. Apocalyptic narratives derived both from Christian tradition and from developments in geological science proved compelling at a time of socio-political upheaval. Disasters and catastrophes tend to be seen as 'rupture[s] or inversion[s] of the normal order of things,'[46] and yet catastrophic thinking may be normalized at certain historical junctures. The texts discussed in this book responded (without fully understanding it) to an extraordinary process—a huge volcanic eruption causing disruption to global climate patterns—but also responded to a more quotidian sense of the catastrophic that pervaded Romantic culture. Given the political and environmental shocks of recent months and years, many of us may share with Romantic writers a sense that 'disaster has already phenomenally occurred within our social, political, and cultural structures.'[47] In the Anthropocene, distinctions between normal and abnormal, order and disorder, are becoming increasingly difficult to make.

Romanticism, Materialism, Nihilism

This book was originally to be focused on the relationship between community and catastrophe. I was interested in how the Tambora crisis affected different communities, in how it produced collaborative writing, and in how that writing imagined new forms of community. However, I subsequently realised that, in the context of writing about human-nonhuman interactions, my idea of 'community' was narrow and anthropocentric. I have therefore increasingly drawn on work that emphasises the complexity of human-nonhuman entanglements and uses different terminology. A key figure is Jane Bennett, whose *Vibrant Matter* has been influential since it was published in 2010. Bennett is concerned with 'the capacity of things—edibles, commodities, storms, metals—not only to impede or block the will and designs of humans but also to act as quasi agents or forces with trajectories, propensities, or tendencies of their own.'[48] Building on Bruno Latour's actor-network theory and the work of Deleuze and Guattari, she develops the idea of the 'assemblage,' a grouping of vibrant materials, which can include human and nonhuman elements, and that manifests distributed agency.[49] Bennett is often connected with other thinkers such as Karen Barad as part of the 'new materialism' that emphasises relational ontologies and the vitality of matter, and which has an important but uneasy relationship to a wider philosophical grouping called speculative realism.[50] This blanket term covers a variety of differing approaches linked together by a concern with the significance of matter beyond its significance for human beings. That is to say, speculative realists critique post-Kantian correlationism: 'the idea according to which we only ever have access to the correlation between thinking and being, and never to either term considered apart from the other.'[51] A key strand of this movement is object-oriented ontology, which emphasises not only the 'realness' of objects but also their ontological withdrawnness from each other and therefore contrasts with what tends to be the more relational theories of the new materialists.[52]

As I discuss in Chap. 3, Timothy Morton draws on object-oriented ontology to develop his analysis of 'hyperobjects' such as global warming: 'things that are massively distributed in time and space relative to humans' and whose 'primordial reality is withdrawn' from us.[53] For Morton, hyperobjects are the final nail in the coffin of anthropocentric thinking.[54] Because speculative realism is concerned with understanding the world separately from human consciousness, it provides an opportunity to reflect on human extinction and can sometimes lead to nihilistic conclusions. As Raymond Brassier argues, there is a sense in which human extinction has already happened because the very thought of it destroys any sense of human transcendence: the idea that the universe only 'means' in so far as we apprehend it.[55] Steven Shaviro suggests that for Brassier, 'the consequence of rejecting correlationism is that we come to recognize a universe that is not only irreducible to thought but fatally inimical to thought.'[56] This book takes from the work of thinkers such as Brassier and Eugene Thacker a sense of the horror of human life within a universe to which it is entirely marginal and meaningless.[57] I find this sense of disenchantment and withdrawal a valuable counter to the rather fuzzy 'everything is connected' mantra sometimes associated with the new materialism, and to the upbeat rhetoric and wishful thinking sometimes associated with the environmental humanities. It is not only that we are at a point now where we need to acknowledge that climate change is a 'wicked problem' that we will have to live with, rather than arrogantly seek to 'solve.'[58] (This acknowledgement, of course, does not preclude doing everything we can to reduce greenhouse gas emissions.) Even more starkly, decentring the human means addressing seriously the possibility, as the bleakest Romantic writing does, that human life is not worth living in the context of a withdrawn and uncaring universe.

The study of Romantic writing should be fertile ground for bringing in new materialist and speculative realist philosophies due to its concern with human-nonhuman interactions and nonhuman agency. And yet until very recently British Romanticism has barely figured in work in this area. There are signs that this situation is changing, most obviously with the publication of Evan Gottlieb's *Romantic Realities*, which explores connections between a wide range of speculative realist philosophies and Romantic poets.[59] While Gottlieb alludes to the value of addressing this relationship at a time of global warming, his study is not in itself an ecocritical one.[60] Indeed, Romanticism is more or less absent from two important recent ecocritical collections that address literature and the new materialism.[61] The study of British Romanticism was integral to the emergence of ecocriticism in the 1980s and 1990s but—with more cutting-edge work taking place in areas such as medieval and postcolonial studies—is in danger of becoming marginal to the growth of material ecocriticism and the environmental humanities more generally.[62] One of the aims of *Writing Tambora*, therefore, is to show how Romantic texts speak to the concerns of this new field and vice versa.

The new materialism has implications not only for the content of our textual analysis, but also for our methodology. Literary discussions of environmental catastrophe have too often been canonically unadventurous, tending to focus on a relatively small (although continually expanding) canon of climate change fiction rather than looking across genres and authors.[63] If we are to understand the ways in which catastrophes are constructed through discourse, we need seriously to address 'non-literary' narratives too, and to place them in juxtaposition with canonical and non-canonical poems, plays, novels, and short stories. Thus my next chapter analyses the heteroglossic government report on the eruption and its local effects as a rhetorical construction. My chapter on the Diodati Circle brings together texts that have previously been linked to Tambora, such as 'Darkness,' *Frankenstein*, and 'Mont Blanc,' with a number that have not, such as *Manfred*, *Ernestus Berchtold*, and *The Prisoner of Chillon*. And my concluding chapter addresses the politics of climate in British periodicals in the post-Tambora period. The focus on distributed agency and networks that is so important to the new materialism might also encourage us to move beyond the fetishisation of individual texts and canonical authors that continues to bedevil literary studies.[64] A more ecocritical approach would be to pay more attention to the materiality of writing itself by considering 'textual assemblages'—interacting groups of texts that produce collaborative meanings—and the significance of these assemblages within, rather than outside, the broader material-cultural assemblages that include environmental processes such as the Tambora eruption. Taking this kind of relational approach puts this study at odds with what Graham Harman describes as 'object-oriented literary criticism,' which he argues should be concerned with how individual literary texts are recalcitrant to contextualisation.[65] However, putting aside the fact that Harman does not acknowledge that 'literature' is a contingent and historically problematic category, I am not convinced that texts can always be usefully understood within the same ontological framework as other sorts of objects. In their case, at least, I am sympathetic to Bennett's response to Harman that 'perhaps there is no need to choose between objects and their relations.'[66]

Climate Change Studies and Anthropocene Historicism

So far, I have introduced two important contexts for my analysis. In studying Tambora as a textual catastrophe, I have drawn on scholarship on the cultural mediation of disaster and on philosophical arguments around material agency. The third of these contexts is climate change studies and particularly what Tobias Menely terms 'Anthropocene historicism.'[67] Timothy Clark has recently argued that the emergence of the Anthropocene calls into question established modes of literary analysis: reading historically with a new consciousness of the complexity of environmental 'scale effects' entails close attention to aspects of a text that may not have been significant when it was produced. Clark understands the Anthropocene as a profound epistemological breach: 'an intractable break in consciousness and understanding.'[68] In this claim, he can be aligned with scholars such as Clive Hamilton and Jacques Grinewald, who have argued that there are no precursors to the idea of the Anthropocene because it is dependent on the relatively recent development of Earth system science and therefore differs in kind from earlier commentary on human environmental impacts.[69] In contrast, Christophe Bonneuil and Jean-Baptiste Fressoz suggest that 'the official narrative of the Anthropocene' makes an unhelpful distinction between 'a blind past and a clear-sighted present,' thereby obscuring the history of environmental reflexivity.[70] The Romantic period is particularly fertile ground for addressing this tension between the Anthropocene as an 'intractable break' and a more genealogical approach. After all, it is often understood not only as the key point of origin for the world's industrialisation and transition into the Anthropocene, but also for modern ideas of ecology. Therefore, when we conduct 'eco-historicist' work on Romantic texts, we need to take account of the fact that those texts may have contributed to the development of our methodology. Indeed, *Writing Tambora* shows how Romantic writers themselves were involved in a kind of 'Anthropocene historicism' by understanding environmental catastrophe in the context not only of geologic time but also of earlier interactions between climate and human culture.

While he is sympathetic to eco-historicism, Clark also has reservations about it:

> This renewed interest in kinds of thought previously overlooked must also avoid the danger of positing the Anthropocene in terms that make it continuous with a long and varied tradition of thought on the relations between weather, climate and human cultures, instead of also being a drastic break with these. Ecocritical reading cannot just be some act of supposed retrieval, but now becomes also a measure of the irreversible break in consciousness and understanding, an emergent unreadability.[71]

There are several grounds on which this argument can be challenged. First, has historicism ever really been about straightforward 'retrieval,' as Clark seems to assume? The best historicist critics have always been aware of a kind of double vision through which the reconstruction of past discursive contexts is informed by present-day concerns. Secondly, one might read the history of ecology and environmental science in the last two centuries not as involving 'a drastic break,' but rather as a gradual intensification of awareness of human impacts, feedback loops, and 'scale effects.' Thirdly, the idea that culture and climate are profoundly intertwined is hardly a new one, and the history of that idea may not be as discontinuous with the Anthropocene as Clark believes. As Mike Hulme puts it, 'the political task is to recognise climate and human life have historically been understood not as two separate domains with one causing or shaping the other. Rather, for much of the past and in most places, climate and humans have been understood to move together, their agency and fate conjoined through the mediating roles of natural processes and supernatural beings.'[72] With all this in mind, there is a danger of overstating the affective and conceptual differences between 'natural' and 'Anthropogenic' climate change as we now understand those terms.[73] Climate change has often been understood to 'move together' with human activity, even as it threatens to overwrite individual agency.[74] Whether seen as natural or human-made, or some combination of the two, an awareness of climate change means that 'Nature' cannot be understood as a fixed backdrop against which human life is played out, but rather as volatile and ever-shifting, revealing the potential limits of our ambitions and our finitude as a species.

As critics, therefore, we need to take care when we jump on the Anthropocene bandwagon. I can certainly see its value as a term of stratigraphic periodisation or as a shorthand for a period of profound human impacts on the Earth system, but I am suspicious of its emergence as a kind of shibboleth in humanities discourse, an apparently easy way of attempting to make our work 'relevant' by suggesting a breach between past and present thinking. I have therefore tried in the present study to invoke it in a nuanced fashion and without flattening the considerable disparities of power that lie at its heart. Part of the problem is an easy slippage between two different uses of the term: (1) a period of time, with a debated start date, during which humans have had profound impacts on the global environment and (2) a mode of thinking about those impacts that is generally associated with present-day reflexivity. As Bonneuil and Fressoz show, such reflexivity actually has a long history; environmental damage has often occurred not because of a lack of understanding of human impacts, but due to political and economic factors. It is certainly the case that Earth system science is a significant intellectual shift, and that in historicising climate change, 'inherited styles of thoughts must be re-examined in a vastly expanded environment of ideas.'[75] But doing so requires that we be more attentive than ever to the complex and sophisticated ways in which past writers themselves understood the relationship between human culture and climate and experienced the blurring of 'distinctions between natural and human history.'[76] If the early nineteenth century saw the gradual attenuation of climatic determinism, Romantic writers remained fascinated by human-climatic interactions.[77] Thus Byron and the Shelleys connect human history with the catastrophic changes of deep time. And Chap. 2 examines how the 'official' narrative of the Tambora eruption places it within a history of connecting political and natural volatility that is shared by colonial and indigenous cultures.

Reading these historical responses to climate change reveals differences from and similarities to our own concerns, and also the sometimes surprising ways in which they present the relationship between human culture and climate. It should be clear by now that this book does not seek to provide a cultural history of Tambora, nor a definitive survey of the literature that it produced. Rather, it uses Tambora as a case study for looking at the cultural history of catastrophe as a discursive-material entanglement. I have written it as a short book, rather than an extended monograph, for two main reasons. The first is timeliness. Although it is convenient that its publication should just about coincide with the bicentenary of the post-Tambora crisis, more importantly I believe that—because of the recent development of material ecocriticism and interest in the concept of the Anthropocene—the intervention that it makes in connecting Romanticism to the environmental humanities will be of particular value now. The second reason is that this book does not mark the end of a project, but the beginning of a larger study of British Romantic literature and environmental catastrophe in relation to our present crisis. It is therefore a kind of experiment: an attempt to discover what Romanticism offers to the environmental humanities and vice versa. My second chapter draws on Bennett's theory of the assemblage to address the heteroglossic narrative of the eruption and its local aftermath that was produced by the British administration in Java. I pay close attention to its publication in three different contexts over a fourteen-year period in order to shed light on the ways in which elemental phenomena and the discourses and practices of imperial and indigenous cultures came together to create 'Tambora.' The following chapter understands the 1816 texts of Byron and the Shelleys as the collaborative products of a creative community and shows that they share a concern with the fragility of human dwelling within a potentially violent universe. In articulating this concern, they draw on a shared language of catastrophe: a kind of textual ecosystem. Their sense of human precarity challenges the reformist and utopian thinking with which Romanticism is often associated. Finally, a concluding chapter considers writing in Romantic periodicals on the 'distresses' afflicting Britain in 1816 and particularly the attacks of reformers on an unjust and exploitative socio-political system. I suggest that understanding the Tambora catastrophe as an 'assemblage' offers a useful case study for addressing questions of human agency and responsibility in the Anthropocene. To return to Blanchot, the disaster threatens to 'lay waste' to human culture, epistemologically and ontologically, and yet also shows how language can 'thrust' being 'out of itself.'[78] To write Tambora, therefore, is to embark on a paradoxical project: to affirm the human capacity to shape and make sense of a world with which we are profoundly entangled, and to accept our humiliation by powerful elemental forces that we do not fully comprehend. I hope that this book does justice to both affirmation and humiliation by showing how writers have responded to the provocations of such a catastrophe with creative energy, as well as horror.

Notes

1. [Charles Assey], 'Narrative of the Effects of the Eruption from the Tomboro Mountain, in the Island of Sumbawa, on the 11th and 12th of April 1815,' *Transactions of the Batavian Society, of Arts and Sciences* 8 (1816): 3–4.
2. [Assey], 23.
3. Clive Oppenheimer, *Eruptions that Shook the World* (Cambridge: Cambridge University Press, 2011), 302–303.
4. For a list of the 25 largest documented Holocene eruptions, see Oppenheimer, *Eruptions*, 356–363.
5. Oppenheimer, *Eruptions*, 311.
6. J. Kandlbauer, P.O. Hopcroft, P.J. Valdes, and R.S.J. Sparks, 'Climate and carbon cycle responses to the 1815 Tambora volcanic eruption,' *Journal of Geophysical Research: Atmospheres* 118 (2013): 12,497–12,507.
7. For the weather conditions in Britain in 1816, see Lucy Veale and Georgina H. Endfield, 'Situating 1816, the "year without summer", in the UK,' *The Geographical Journal*: 182 (2016): 318–330.
8. John D. Post, *The Last Great Subsistence Crisis in the Western World* (Baltimore and London: The Johns Hopkins University Press, 1977).
9. The best study is Gillen D'Arcy Wood, *Tambora: The Eruption that Changed the World* (Princeton: Princeton University Press, 2014). Other valuable accounts include Bernice de Jong Boers, 'Mount Tambora in 1815: A Volcanic Eruption in Indonesia and its Aftermath,' *Indonesia* 60 (October 1995): 37–59; William K. Klingaman and Nicholas P. Klingaman, *The Year Without A Summer: 1816 and the Volcano that Darkened the Sky and Changed History* (London: St. Martin's Press, 2013); Oppenheimer, *Eruptions*, Chap. 13.
10. For an admirably lucid and insightful account of the emergence of the Anthropocene and its political significance, see Jeremy Davies, *The Birth of the Anthropocene* (Oakland: University of California Press, 2016).

11. See, for example, Andreas Malm and Alf Hornborg, 'The geology of mankind? A critique of the Anthropocene narrative,' *The Anthropocene Review* 1 (2014): 62–69.
12. Stacy Alaimo, 'Elemental Love in the Anthropocene,' in Jeffrey Jerome Cohen and Lowell Duckert, eds., *Elemental Ecocriticism: Thinking with Earth, Air, Water, and Fire* (Minneapolis and London: University of Minnesota Press, 2015), 298–309 (307).
13. See, for example, Clive Hamilton, 'The Anthropocene as rupture,' *The Anthropocene Review* 3 (2016): 93–106.
14. The latter date of around 1950 seems the most likely one to be taken up by the International Commission on Stratigraphy.
15. See, for example, Jan Zalasiewicz, Mark Williams, et al., 'Are we now living in the Anthropocene,' *GSA Today* 18 (2008): 4–8, and Victoria C. Smith, 'Volcanic markers for dating the onset of the Anthropocene,' in C. N. Waters, J.A. Zalasiewicz, M. Williams, M.A. Ellis, and A.M. Snelling (eds.), *A Stratigraphical Basis for the Anthropocene* (London: Geological Society Special Publications, 2013): 283–299. For an overview of the various starting dates that have been suggested, see Bruce D. Smith and Melinda A. Zeder, 'The onset of the Anthropocene,' *Anthropocene* 4 (2013): 8–13.
16. Timothy Clark, *The Cambridge Introduction to Literature and the Environment* (Cambridge: Cambridge University Press, 2011), 11.
17. Thus a search of the MLA International Bibliography for ecocritical work, using the terms 'climate change,' 'ecocriticism' AND 'climate change,' and 'literature' AND 'climate change' reveals: (i) sporadic results before 2008; (ii) 16 articles, 2 journal special issues, 2 book chapters, and 2 books between 2008 and 2011; (iii) 46 articles, 4 journal special issues, 16 book chapters, and 9 books between 2011 and 2015. (I am grateful to Tess Somervell for this information.)
18. See, for example, Wolfgang Behringer, *A Cultural History of Climate* (Cambridge: Polity, 2009) and Tom Bristow and Thomas H. Ford, eds., *A Cultural History of Climate Change* (London: Routledge, 2016).
19. Wood, *Tambora* and Klingaman and Klingaman, *The Year Without A Summer*. Other works on the history of climate change and catastrophe, aimed at a wide audience, include: Brian Fagan, *The Little Ice Age: How Climate Made History 1300–1850* (New York: Basic, 2000); Richard Hamblyn, *Terra: Tales of the Earth* (London: Picador, 2009); Geoffrey Parker, *Global Crisis: Climate Change and Catastrophe in the Seventeenth Century* (New Haven and London: Yale University Press, 2013); and Alexandra Witze and Jeff Kanipe, *Island on Fire: The Extraordinary Story of Laki, the Volcano that Turned Eighteenth-Century Europe Dark* (Place: Profile, 2014).

20. Jonathan Bate, *The Song of the Earth* (London: Picador, 2000), Chap. 4; Alan Bewell, *Romanticism and Colonial Disease* (Baltimore and London: The Johns Hopkins University Press, 1999), 223–226.
21. Serenella Iovino and Serpil Oppermann, 'Introduction,' in *Material Ecocriticism*, ed. Serenella Iovino and Serpil Oppermann (Bloomington and Indianapolis: Indiana University Press, 2014), 9. For a fine recent example of material ecocriticism that brings together book history and geoscience, see Eric Gidal, *Ossianic Unconformities: Bardic Poetry in the Romantic Age* (Charlottesville and London: University of Virginia Press, 2015).
22. See, for example, its use by Oppenheimer, *Eruptions*, Chap. 13.
23. For an interesting reading of Blanchot's book that acknowledges its lack of interest in ecology while placing it in relation to the Anthropocene, see Joshua Schuster, 'How to Write the Disaster,' *Minnesota Review* 83 (2014): 163–171.
24. Maurice Blanchot, *The Writing of the Disaster*, trans. Ann Smock (Lincoln and London: University of Nebraska Press, 1995), 7.
25. Blanchot, *Writing*, 99.
26. Blanchot, *Writing*, 7.
27. Blanchot, *Writing*, 99.
28. Blanchot, *Writing*, 124.
29. Gilles Deleuze and Félix Guattari, *A Thousand Plateaus: Capitalism and Schizophrenia*, trans. Brian Massumi (London: Bloomsbury, 2016), 10.
30. Mark D. Anderson, *Disaster Writing: The Culture Politics of Catastrophe in Latin America* (Charlottesville and London: University of Virginia Press, 2011), 1.
31. Anderson, *Disaster Writing*, 1.
32. See, for example, Anthony Carrigan, '"Out of this great tragedy will come a world class tourism destination": Disaster, Ecology, and Post-Tsunami Tourism Development in Sri Lanka,' in *Postcolonial Ecologies: Literatures of the Environment*, ed. Elizabeth DeLoughrey and George Handley (Oxford: Oxford University Press, 2011), 273–290; Nigel Clark, *Inhuman Nature: Sociable Life on a Volatile Planet* (London: Sage, 2011), Chaps 3 and 6; Sharon Monteith, ed., *Hurricane Katrina: Five Years After*, special issue of the *Journal of American Studies* 44 (2010).
33. Anderson, *Disaster Writing*, 7.
34. Anthony Carrigan, 'Towards a Postcolonial Disaster Studies,' in *Global Ecologies and the Environmental Humanities: Postcolonial Approaches*, ed. Elizabeth DeLoughrey, Jill Didur, and Anthony Carrigan (London: Routledge, 2015), 121.
35. Carrigan, 'Postcolonial Disaster Studies,' 123.
36. Rob Nixon, *Slow Violence and the Environmentalism of the Poor* (Cambridge, MA: Harvard University Press, 2011), 32.

37. Timothy Clark, *Ecocriticism on the Edge: The Anthropocene as a Threshold Concept* (London: Bloomsbury, 2015), 109, 72.
38. Kate Rigby, *Dancing with Disaster: Environmental Histories, Narratives, and Ethics for Perilous Times* (Charlottesville and London: University of Virginia Press, 2015), 15.
39. Rigby, *Dancing*, 22.
40. I made this decision before reading Rigby's somewhat similar reasoning for preferring the term '*eco-catastrophe*' (Rigby, *Dancing*, 17–18).
41. Rigby, *Dancing*, 18.
42. Timothy Morton, 'Romantic Disaster Ecology: Blake, Shelley, Wordsworth,' in *Romanticism and Disaster*, eds. Jacques Khalip and David Collings, *Romantic Circles* (January 2012), paragraph 7, accessed 22 September 2016, http://www.rc.umd.edu/praxis/disaster/HTML/praxis.2012.morton.html.
43. Morton, 'Romantic Disaster Ecology,' paragraph 5.
44. See Timothy Morton, *The Ecological Thought* (Cambridge, MA: Harvard University Press, 2008).
45. Morton, 'Romantic Disaster Ecology,' paragraph 7.
46. Anderson, *Disaster Writing*, 1.
47. Jacques Khalip and David Collings, 'Introduction: The Present Time of "Live Ashes",' in *Romanticism and Disaster*, *Romantic Circles* (January 2012), paragraph 10, accessed 22 September 2016, http://www.rc.umd.edu/praxis/disaster/HTML/praxis.2012.khalip.html.
48. Jane Bennett, *Vibrant Matter: A Political Ecology of Things* (Durham and London: Duke University Press, 2010), viii.
49. Bennett, *Vibrant Matter*, Chap. 2. For a critique of the idea of 'distributed agency,' see Alf Hornborg, 'Artifacts have consequences, not agency: Toward a critical theory of global environmental history,' *European Journal of Social Theory* 20 (2016): 1–16. For a different approach, which emphasises the 'utter confusion between objects and subjects' in the Anthropocene, see Bruno Latour, 'Agency at the Time of the Anthropocene,' *New Literary History* 45 (2014): 1–18.
50. Karen Barad, *Meeting the Universe Halfway: Quantum Physics and the Entanglement of Matter and Meaning* (Durham and London: Duke University Press, 2007) and 'Posthumanist Performativity: Towards an Understanding of How Matter Comes to Matter,' *Signs: Journal of Women in Culture and Society* 28 (2003): 801–831. For a helpful critical overview of speculative realism, see Steven Shaviro, *The Universe of Things: On Speculative Realism* (Minneapolis: The University of Minnesota Press, 2014).
51. Quentin Meillassoux, *After Finitude: An Essay on the Necessity of Contingency*, trans. Ray Brassier (London: Bloomsbury, 2008), 5.

52. The key thinker here is Graham Harman: see, for example, *Guerrilla Metaphysics: Phenomenology and the Carpentry of Things* (Chicago: Open Court, 2005) and 'The Well-Wrought Broken Hammer: Object-Orientated Literary Criticism,' *New Literary History* 43 (2012): 183–203. Jane Bennett responds to Harman's critique of relational ontologies in 'Systems and Things: A Response to Graham Harman and Timothy Morton,' *New Literary History*, 43 (2012): 225–233.
53. Timothy Morton, *Hyperobjects: Philosophy and Ecology After the End of the World* (Minneapolis: University of Minnesota Press, 2013), 1, 15.
54. Morton, *Hyperobjects*, 18–19.
55. Obviously this single sentence can only offer a very crude summary of Brassier's complex argument: see *Nihil Unbound: Enlightenment and Extinction* (Basingstoke: Palgrave, 2007), especially Chap. 7.
56. Shaviro, *Universe*, 122.
57. Brassier, *Nihil Unbound*, and Eugene Thacker, *In The Dust of this Planet: Horror of Philosophy, Vol. 1* (Winchester: Zero Books, 2010).
58. See Roy Scranton, *Learning to Die in the Anthropocene: Reflections on the End of a Civilisation* (San Francisco: City Lights Books, 2015), Chap. 2; Dipesh Chakrabarty, 'Humanities in the Anthropocene: The Crisis of an Enduring Kantian Fable,' *New Literary History* 47 (2016): 377–397 (383); Mike Hulme, *Why We Disagree About Climate Change: Understanding Controversy, Inaction, and Opportunity* (Cambridge: Cambridge University Press, 2009), 334–337.
59. Evan Gottlieb, *Romantic Realities: Speculative Realism and British Romanticism* (Edinburgh University Press, 2016). See also Greg Ellermann, 'Speculative Romanticism,' *SubStance* 44.1 (2015): 154–174 and Chris Washington, 'Romanticism and Speculative Realism,' *Literature Compass* 12.9 (2015): 448–460.
60. Gottlieb, *Romantic Realities*, 234.
61. Cohen and Duckert's *Elemental Ecocriticism* takes an ecomaterialist approach that emphasises 'environmental agentism' and argues that we should 'explore the dis-anthropocene' and 'dive deeper into our elemental embeddedness' (6, 18). Romantic writing is often concerned with that embeddedness, and yet it barely figures in the book. Similarly, Iovino and Opperman's *Material Ecocriticism*—which examines 'matter both *in* texts and *as* a text' (2)—contains only passing references to British Romantic authors. I point to these lacunae not to criticise these valuable studies, but to draw attention to the fact that Romanticism has not figured strongly in the development of this important new area of research.
62. For Romanticism and early ecocriticism see, for example, Jonathan Bate, *Romantic Ecology: Wordsworth and the Environmental Tradition* (London: Routledge, 1991); Karl Kroeber, *Ecological Literary Criticism: Romantic*

Imagining and the Biology of Mind (New York: Columbia University Press, 1994); James C. McKusick, *Green Writing: Romanticism and Ecology* (Basingstoke: Palgrave, 2000).

63. Thus *The Last Man* has produced numerous recent ecocritical readings. While it is a fascinating novel, we may be in danger of diminishing analytical returns from neglecting the many other texts from the period that deal with environmental catastrophe.
64. Clearly there is a growing body of work on literary networks and sociality that goes against this tendency, including recent studies such as Tim Fulford, *Romantic Poetry and Literary Coteries: The Dialect of the Tribe* (New York: Palgrave, 2015) and Jon Mee, *Conversable Worlds: Literature, Contention and Community, 1762–1830* (Oxford: OUP, 2011). However, the discourse of individual genius that was to some extent developed during the Romantic period remains remarkably persistent.
65. Harman, 'The Well-Wrought Broken Hammer.'
66. Bennett, 'Systems and Things,' 227.
67. I have taken this term from the title of the special seminar organised by Professor Menely at the North American Society for the Study of Romantic conference in Berkeley in 2016.
68. Clark, *Ecocriticism on the Edge*, 54.
69. Clive Hamilton and Jacques Grinewald, 'Was the Anthropocene anticipated?,' *The Anthropocene Review* 2 (2015): 59–72.
70. Christophe Bonneuil and Jean-Baptise Fressoz, *The Shock of the Anthropocene*, trans. David Fernbach (London: Verso, 2016), xii–xiii. Alan Mikhail takes a similar view, noting that 'the Anthropocene narrative' borrows much from 'the Enlightenment narrative,' and both assume 'too clean a break between the modern age and everything before it': see Alan Mikhail, 'Enlightenment Anthropocene,' *Eighteenth-Century Studies* 49 (2016): 211–231 (226).
71. Clark, *Ecocriticism on the Edge*, 62. Clark refers to *Eco-Historicism*, a special issue of *The Journal for Early Modern Cultural Studies*, ed. Gillen D'Arcy Wood, 8 (2008).
72. Mike Hulme, *Weathered: Cultures of Climate* (London: Sage, 2017), 50.
73. The terms 'Anthropogenic'/'Anthropocene' can obviously be problematic if they form part of a discourse that overstates human agency and control over an externalised 'environment.' Even 'Anthropogenic' climate change results from the actions of an assemblage of agents, not just humans. Nigel Clark addresses the challenges that natural volatility poses for social theory in *Inhuman Nature*.

74. As Dipesh Chakrabarty puts it, the Anthropocene implies collective human agency but 'there could be no phenomenology of us as a species': 'The Climate of History: Four Theses,' *Critical Inquiry* 35 (2009): 197–222 (220). See also his article 'Postcolonial Studies and the Challenge of Climate Change,' *New Literary History* 43 (2012): 1–18.
75. Bristow and Ford, *Cultural History*, 7.
76. Bristow and Ford, *Cultural History*, 3.
77. For climatic determinism, see Jan Golinski, *British Weather and the Climate of Enlightenment* (Chicago and London: The University of Chicago Press, 2007), Chap. 6.
78. Blanchot, 99, 124.

CHAPTER 2

Textuality, Empire, and the Catastrophic Assemblage: Sir Stamford Raffles and the Tambora Eruption

Abstract This chapter focuses on the English-language account of the Tambora eruption produced under the auspices of the British Governor of Java, Sir Stamford Raffles. Scholars have treated this text as a straightforward account of the eruption and its effects, but it is in fact a complex heteroglossic and collaborative production in which different perspectives and knowledges intertwine and compete. 'Tambora', the chapter argues, is usefully understood as a material-discursive assemblage in which the natural and the political are thoroughly imbricated. These imbrications are addressed through a close analysis of Raffles's narrative in its three different contexts of publication over a fifteen-year period. The chapter shows, in particular, how the document supplants a mythic understanding of catastrophe with an imperalistic, 'scientific' epistemology.

Keywords Material ecocriticism · New materialism · British imperialism · Sir Stamford Raffles · Tambora

Introduction

Due to the short interregnum from 1811 to 1816 during which Britain wrested control of Java from Holland—which in 1811 was part of the French empire—the principal source for eyewitness accounts of the Tambora eruption and its aftermath is an English-language document

D. Higgins, *British Romanticism, Climate Change, and the Anthropocene*, DOI 10.1007/978-3-319-67894-8_2

collected together under the auspices of Sir Stamford Raffles, the island's governor during this period, and published in the *Transactions of the Batavian Society* in 1816.[1] Scholars have tended to treat this text as a straightforward source of information about the eruption and its effects. However, it is in fact a heteroglossic and collaborative production in which different forms of knowledges intertwine.[2] Raffles's desire to mediate and control representations of the eruption within a single centralising document was connected to his role as a colonial administrator: throughout his time in Java he was gathering information to use in the governance of the archipelago and to make a case for a permanent British colonial presence there, as he does in the *History of Java* (1817). There is a tension between the imperialistic and bureaucratic metanarrative provided by Raffles and his colonial functionaries, and the text's more localised accounts of the catastrophe. Although the report emphasises the sublime power of the eruption, it also presents itself as a repository of objective environmental knowledge that will support imperial power and control. In contrast, the individual accounts that it also contains are more often shown to invoke confusion, legend, and ignorance. The document therefore can be seen as marking a shift from a religious-mythic understanding of disaster in which 'morality and materiality, social relations and natural phenomena, were understood to be interrelated' to a more 'modern understanding [...] according to which the merely material realm of nature followed its own mechanistic principles that were entirely separate from human morality and social relations.'[3] It reveals a nascent and uneven distinction between a 'fact-making,' supposedly objective and scientific worldview embodied by the colonial metanarrator and the more embedded 'meaning-making' individual narratives within it.[4]

It is fitting that the key source of information about the Tambora eruption is a collection of different perspectives. The catastrophe itself, after all, was not a single event. To use Jane Bennett's term, it was an 'assemblage': a grouping 'of diverse elements, of vibrant materials of all sorts.'[5] The word 'Tambora' can stand simply for the volcano itself, but it can also be used to refer to this assemblage of energy and matter, which includes the initial explosions, the Plinian eruption, the lava flow, the pyroclastic currents that destroyed villages, the ash fallout, the darkness that covered a large area, the fall of pumice from the sky, the tsunami, and even the subsequent global climate change caused by the release of sulphur into the stratosphere. Bennett argues that the elements that make up an assemblage can include 'humans and their (social, legal, linguistic) constructions,' as is made evident by her case study of the electrical power grid.[6] An assemblage, too, can have a form of agency due to 'the vitality of the materials that constitute it.'[7] Bennett's work and that of other new materialist thinkers thereby problematizes the dichotomy by which human beings are seen either as shapers of nonhuman nature or victims of its whims by focusing instead on 'a material-semiotic network of human and nonhuman agents incessantly generating the world's embodiments and events.'[8] An uncomfortable question raised by a phenomenon like Tambora is the extent to which humans and animals participate within it rather than simply being acted upon by it. To make such a suggestion in the face of the immense suffering that accompanies an environmental catastrophe on such a scale may seem like a kind of obscenity. I write this paragraph ten days after an earthquake devastated towns and villages in central Italy, killing nearly 300 people and injuring many more. It is unlikely that any of those people or their loved ones thought of themselves as participants in an assemblage. And yet an earthquake does not in itself kill anyone. What we understand as an environmental catastrophe is a set of interactions between earth systems, organic beings, and their discursive-material constructions.

Bennett's emphasis on the vitality of matter, and the potential inclusion of human beings within assemblages, is why I have leant towards her terminology in this chapter rather than Timothy Morton's more distancing concept of the 'hyperobject': a thing that is 'massively distributed in time and space relative to humans.'[9] However, 'Tambora' certainly manifests the nonlocality, viscosity, and interobjectivity that Morton sees as among the key properties of such objects. In any case, both concepts address the potentially asymmetrical entanglements of human beings and powerful natural forces that the Tambora narrative documents. It also shows how individual human experience of a large assemblage or hyperobject is inevitably partial and mediated, as evidenced by the silent darkness that descends across a large area after the eruption: an experience that is simultaneously revealing and obscuring. The document's metanarrative attempts to overcome individual partiality by bringing together different perspectives, and correcting them when necessary, but itself inevitably offers a partial view. Most importantly, as a document of empire in the language of the invading country it tends to obscure indigenous voices, and indigenous people bore the brunt of the volcano's effects. The few indigenous perspectives that we do encounter are mediated by being reported within the accounts of European witnesses and within the document's metanarrative. Whereas we might expect a fictionalised account of such an event to attempt to evoke empathy with the suffering of affected people, perhaps through depictions of interiority, the narrative is mainly concerned with the observations of colonising subjects, made from a position of relative safety, whether they are of unusual natural phenomena, or of the destruction that these phenomena wreak. The apparently objective and empiricist mode of the document is no doubt why it has been so useful for scholars trying to uncover the history of the eruption, and a basic reading can tell us a great deal about the different physical phenomena that made up the Tambora assemblage. But a more sophisticated analysis needs to recognise that there is also much that it does not tell us, particularly about the human costs of the eruption. The document offers, via its metanarrative, a particular perspective that encodes a kind of cultural politics. Therefore I approach the narrative not so much as an external account of Tambora, but as a discursive construction that is itself an insertion into the assemblage, and I pay considerable attention to the different contexts in which it was published.

As I discussed in the introduction, this approach builds on recent work in 'material ecocriticism,' which emphasises the 'constitutive engagement of human discursive systems with the material world.'[10] I draw not only on Bennett's idea of the assemblage, but also (selectively) on its rather different conceptualisation by Deleuze and Guattari, who are more sensitive to the processual nature of assemblages.[11] They understand an assemblage as 'simultaneously and inseparably a machinic assemblage and an assemblage of enunciation'; that is, material and discursive.[12] An assemblage inevitably marks out a territory but is also 'constituted by *lines of deterritorialization* that cut across it and carry it away.'[13] The Tambora assemblage, therefore, can be understood not only as a physical process but also as a discursive one in which a mythic understanding of environmental catastrophe, which might include European as well as indigenous epistemologies, is to some extent supplanted (deterritorialized and then reterritorialized) by an imperialist, 'fact-making' epistemology. To take this approach is not to downplay the terrible human effects of 'Tambora'; quite the contrary. Focusing on its discursive as well as physical aspects offers a way of thinking about how environmental catastrophes imbricate the natural and the political.

The first part of this chapter reveals the constructed nature of Raffles's narrative by analysing it as a composite text and paying close attention to its status as the production both of an imperial bureaucracy and of the Batavian Society of Arts and Sciences. Part two analyses what the document reveals about the eruption's 'nonlocality' in relation to the aspirations of British imperialism, before analysing the most direct account that we have of the eruption and its devastating local effects, in Owen Phillips's report that appears at the end of the narrative. In parts three and four, I consider the Tambora assemblage as a continuing process of imperialistic reterritorialization after the initial 1816 publication of the narrative. I focus particularly on its appearance in the form of a lengthy footnote to an early chapter of Raffles's *History of Java* (1817) and in his wife's posthumous *Memoir* of 1830.[14] An analysis of these later publications reveals some important political resonances in Raffles's account, and I focus particularly on the relationship between Tambora and the catastrophic defeat of Napoleon in the context of British imperialism after Waterloo. If material ecocriticism requires close attention to 'the intra-action of human creativity and the narrative agency of matter,' a crucial locus for such attention is the materiality of the text itself and the changing history of that materiality.[15] The Tambora narrative's manifestation in different publishing contexts sheds light on the ways in which elemental phenomena and the discourses and practices of imperial and indigenous cultures came together to create Tambora.[16]

The Original Narrative: Context and Composition

The original document on the Tambora catastrophe is around 4500 words in length and entitled 'Narrative of the Effects of the Eruption from the Tomboro Mountain, in the Island of Sumbawa, On the 11th and 12th of April 1815.' It is dated 28 September 1815, and was first published in the eighth volume (1816) of the Anglo-Dutch *Transactions of the Batavian Society, of Arts and Sciences.*[17] The narrative is unsigned but is introduced as '*communicated by the [Society's] President.*' Introducing the version that appears in the *History of Java*, Raffles notes that it was 'drawn up by Mr. [Charles] Assey' from the responses to Raffles's circular requiring information about the eruption. Assey was at the time the Secretary to the British Government in Java.[18] The narrative is sometimes cited as authored by J.T. Ross, although I am not aware that any scholar has discussed the attribution.[19] The *Java Half-Yearly Almanac and Directory for 1815* lists the Rev. Professor J.T. Ross as minister of the Protestant Church and connected to the government administration in Batavia.[20] The eighth volume of the *Transactions of the Batavian Society* also lists him as the Society's President—he had previously been its Secretary—as Raffles had stepped down at some point after learning in September 1815 that the East India Company was dismissing him from his post and that he would be leaving Java. (The volume begins with an expression of gratitude to Raffles from the Society's post holders, followed by Raffles's response.) Whoever first cited the narrative as by Ross seems to have assumed that '*communicated by the President*' indicates that he was the author. However, given the date of the document, which is well before Raffles left Java on 25 March 1816, it seems much more likely that the reference is to the Society's outgoing president rather than its incoming one.[21] And, even if it does refer to Ross, '*communicated*' is likely to mean no more than it was delivered by him at a Society meeting. There is absolutely no reason to think that Ross had anything to do with its composition or to doubt Raffles's later claim that it was put together by Assey. Previous scholars have not troubled themselves about what may seem a pedantic point.[22] However, the ambiguity about the document's authorship tells us something about its collaborative nature and its bureaucratic stance: a text that is 'drawn up' rather than composed or created.[23] Even as it draws on subjective accounts, this impersonality supports the narrative's claim to authoritative and supposedly objective 'fact-making.'

To understand the document fully, it is useful to know something about the Society. With his customary energy, Raffles sought to find out as much as possible about Java once he had become Lieutenant-Governor in late 1811. As well as travelling the island himself, he commissioned researchers such as the natural historian Thomas Horsfield. He also sought to revive the Batavian Society of Arts and Sciences, which had been founded by the Dutch in 1778, but had become moribund due to the indolence and in-fighting of its members.[24] New bylaws were passed in November 1812 and Raffles was elected President in April 1813. Like its Dutch forebear, this new version of the Society had as its principal objective the pursuit of knowledge for the 'public benefit' and, although obviously a part of the colonising process, was not designed to undertake business for the government or the East India Company.[25] Nonetheless, Raffles held the Society's meetings at his government residence and printed its *Transactions* freely using the administration's press, so it had at least a semi-official imprimatur and noted in its new regulations that it was 'under the special patronage of the British

Government.'[26] Raffles clearly felt that a greater understanding of the natural history of Java and surrounding islands, and of the indigenous languages, would help to develop Java as a productive colony to the benefit of colonisers and colonised:

> As every untrodden path affords some new incitement to the inquisitive mind, so we may look for much in the various branches of Natural History;—to the philosophic mind a wide and interesting field is opened, and while we contemplate in a political point of view the advantages which must result from bringing forth and directing in a proper course the latent energies and resources of so large a portion of the habitable globe, it must be a pleasing reflection to the Philanthropist that so many of our fellow creatures are thus gradually retrieved from ignorance, barbarism, and self destruction.[27]

There could hardly be a clearer articulation of the imperial ideal of a beneficial relationship between scientific knowledge, political power, colonial development, and moral improvement. It assumes, however, that the area's 'latent energies' are capable of being properly directed; one wonders if the Tambora eruption gave Raffles pause for thought about the potential recalcitrance of the Indonesian environment and its peoples to colonial 'improvement.'

Two volumes of the Society's transactions were published during the period of British rule. Volume VII appeared in 1814, the first for twenty-two years, and Volume VIII in 1816. A handful of the articles were in Dutch but the majority were in English: mostly on natural history or philology. The Tambora narrative therefore fitted in very well with the journal and with the Society's mission of increasing Western knowledge of the archipelago. However, unlike the other articles, it was also an official government document. Figure 2.1 gives a visual representation of the text's structure. It comprises an introduction, and then extracts from letters by the Residents (East India Company officials) at Grissie, Sumanap, Baniowangie, and Bima, an unnamed person based at Fort Malboro on Sumatra, and from a Lieutenant Owen Phillips, the Assistant Resident at Macassar, whom Raffles dispatched to Sumbawa on a relief mission. It also contains accounts from 'the Noquedah [Captain] of a Malay Prow' (mediated in the third person) passing near Tambora on 11 April, the commander of 'the Honorable Company's Cruizer Benares' (first person), and the 'Honorable Company's Cruizer

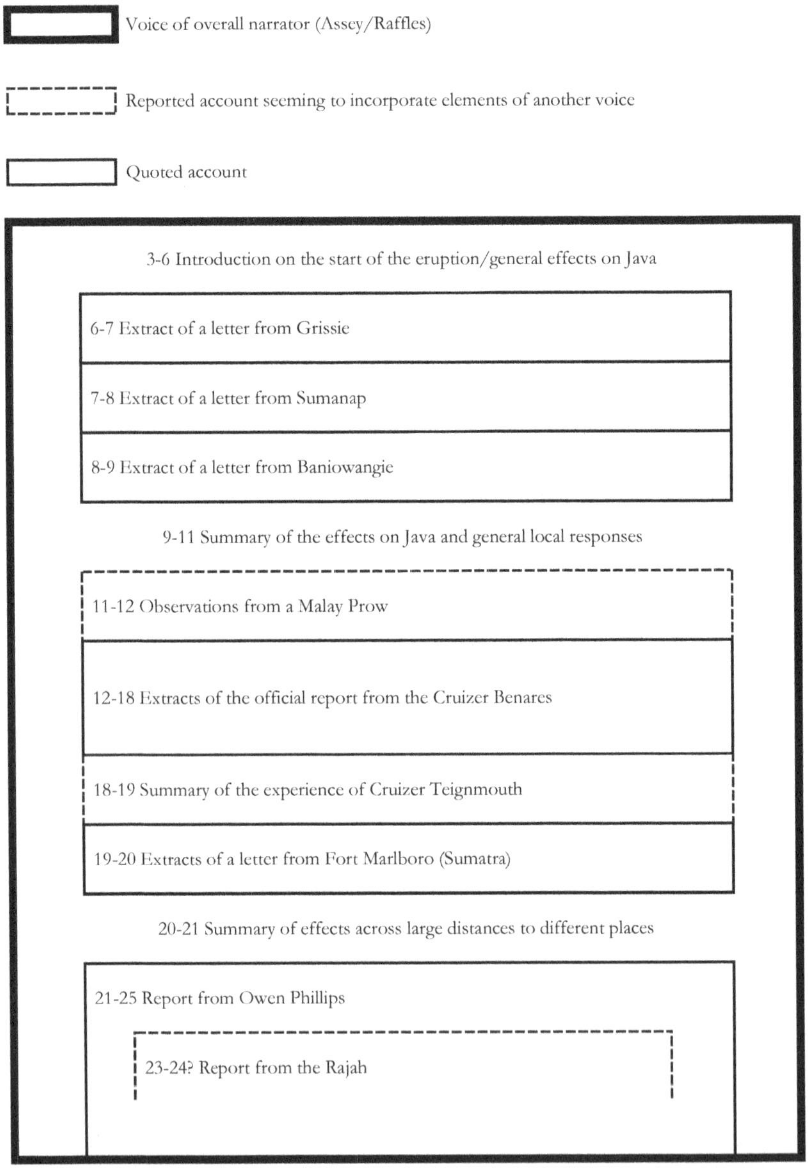

Fig. 2.1 Diagram of the 'Narrative of the Effects of the Eruption from the Tomboro Mountain'

Teignmouth' (third person). Within these narratives are general comments on indigenous responses as well as a 'direct' eyewitness account of the eruption by the Rajah of Saugar that is incorporated within Phillips's narrative. On the one hand, this elaborate structure might suggest the impossibility of offering a single true rendering of any catastrophe. On the other, the implication is that, through its distanced perspective, the imperial centre has the capacity to harness these different local accounts and therefore to produce authentic and stable knowledge. This centralising, territorializing impetus works against the nonlocality of the Tambora assemblage: its tendency to disperse outwards from the site of the eruption. Colonial fact-making is also built on a vast lacuna, which the Rajah's brief and heavily mediated account cannot fill: the meaning-making stories of the indigenous inhabitants of Sumbawa and other islands who experienced the eruption not as a sublime curiosity, but as an immediate threat to life and livelihood.

The opening of the narrative is brisk and functional, and establishes the document's status as the collaborative product of an imperial bureaucracy rather than of an individual observer:

> To preserve an authentic account of the violent and extraordinary Eruption of the Tomboro Mountain on Sumbawa in April last, the Honorable the Lieutenant Governor required from the several Residents of districts on the Island, a statement of the circumstances that occurred within their knowledge, and from their replies the following Narrative is collected. (3)

The 'extraordinary' violence of the eruption is brought into the control of the machinery of colonial government; the word 'required' emphasises that this is knowledge produced by centralised power. The passive voice at the end of the quotation ('is collected') suggests the impersonality of this process. As a result, the narrative is able to offer an overview of the eruption and its effects that supposedly transcends the limitations of the peripheral and localised knowledge offered by the Residents: it can therefore claim to be 'an authentic account.'[28] The entirely ordinary administrative action displayed by the Governor's directions, the obedient responses of the Residents, and the production and publication of the narrative is produced by the very extraordinariness of the events described. The only way to deal with the catastrophic, the document implies, is to be as measured and orderly as possible.

A notable feature of the document is its listing of different local explanations for the noise of the eruption and for the eruption itself. It is clear not just from the narrative, but from other accounts printed in periodicals, that the immediate assumption of anyone hearing the explosions without experiencing any of the other phenomena such as the ash fall and darkness was that guns were being fired. As Raffles's document states, on Java 'the noise was in the first instance almost universally attributed to distant cannon; so much so that a Detachment of Troops were marched from Djocjokarta, in the expectation that a neighbouring Post was attacked, and along the Coast boats were in two instances dispatched in quest of a supposed ship in distress' (3–4). Cannon-like sounds were also heard in many other places, including Macasser on South Sulawesi: 'towards sun-set the reports seem to have approached much nearer and sounded like heavy guns with occasional slight reports in between' (12–13). Assuming pirate activity, 'a Detachment of Troops was embarked on board the Honorable Company's Cruizers Benares and sent in search of them' (13). (These accounts are a reminder that the British operation in Indonesia was very much on a military basis and that its troops were constantly on the alert for pirates, as well as a Dutch invasion.) Even when the fall of ash reveals that a volcano is the source of the noises, the document emphasises that nobody realises that the eruption is as far away as Sumbawa: 'it was attributed to an Eruption from the Marapi, the Gunung Kloot or the Gunung Bromo' (4).

These colonial misunderstandings of the source of the noise are reported alongside indigenous readings of the eruption. The Resident at Grissie (in East Java) relates the following response to the darkness and ash fall:

> I am universally told that no one remembers, nor does their tradition record, so tremendous an Eruption—some look upon it as typical of a change, of the re-establishment of the former Government; others account for it in an easy way by reference to the superstitious notions of their legendary tales, and say that the celebrated Nyai Lorok Kidul has been marrying one of her children, on which occasion she has been firing salutes from her supernatural Artillery. They call the ashes the dregs of her Ammunition. (7)

These readings of the eruption as symbolising political or supernatural events contrast strongly with the scrupulously empirical account by the

Resident himself (discussed below). The phrase 'superstitious notions' suggests his dismissive attitude to such narratives. A little later in the document, the metanarrator reports more neutrally that 'the Balinese attributed the event to a recent dispute between the two Rajahs of Bali B'liling, which terminated in the death of the younger Raja by order of his Brother' (9–10). At this point in the document, before the eruption has been discussed in any detail, the indigenous interpretations do not seem any more or less valid than the more rationalistic reports that also misunderstand the eruption's source. The implication of the Raffles account is that the true story of the catastrophe can only emerge from this *post-facto*, totalising, and disinterested narrative which has the capacity to contain a variety of responses and accounts.

Many cultures have interpreted unusual natural events such as volcanic eruptions as presaging or reflecting a likely political or social change. In the case of Tambora, the inhabitants of east Java were in fact correct that the 'former Government' would shortly be re-established. Given that the narrative is dated 28 September 1815, we might reasonably assume that it was composed during that month. During 1814 and 1815, the Raffles administration existed in an unstable geopolitical context. Java had been returned to the Dutch by the Anglo-Dutch Treaty of August 1814. A year later, news arrived there of Napoleon's escape and resurgence, which potentially voided the Treaty and excited Raffles with the possibility of a longer British presence on the island. And yet in September he received the devastating news that he was likely to be dismissed from office. That this sense of political instability may have informed the Tambora narrative is apparent in its description of the weather on Java following the initial explosions:

> From the 6th, the sun became observed: it had every where the appearance of being enveloped in fog, the weather was sultry and the atmosphere close and still; the sun seemed shorn of its rays, and the general stillness and pressure of the atmosphere foreboded an Earthquake. (4)

This description alludes, rather gracefully, to the famous passage in Book 1 of *Paradise Lost* describing Satan as a 'ruin'd' archangel after his fall:

> As when the Sun new-ris'n
> Looks through the Horizontal misty Air
> Shorn of his Beams, or from behind the Moon

In dim Eclipse disastrous twilight sheds
On half the Nations, and with fear of change
Perplexes Monarchs.[29]

Allusions to Milton were conventional in descriptions of the sublime in the period.[30] But this is a particularly apt one. It connects the mist-covered sun to a catastrophic overturning of the normal state of things—as indeed it was—and, more specifically, it connects meteorological phenomena to political ones.[31] While emphasising the status of the document as a production of elite Western culture, the allusion also connects the imperialistic metanarrative to the more localised indigenous accounts that read sudden environmental change as signalling some change in the state of human affairs. It also draws attention to the complex mediations that inform the narrative. Individual experiences of the eruption are shown to be mediated through distance, fog, and darkness; eyewitness accounts are mediated through the metanarrator; the metanarrator's perspective is here mediated through the Miltonic sublime; and the narrative is itself mediated through Assey, Raffles, and the different contexts of its publication.[32] The narrative's claims to exert discursive authority from a distanced perspective are to some extent contradicted by its heteroglossic mode, and a complex entanglement of stories that are all shown to be part of the Tambora assemblage, rather than external to it.

The Original Narrative: Placing Tambora

The narrative shows that the Tambora catastrophe occurs not just in the immediate area of the eruption, but across the archipelago. An analysis of the eyewitness accounts from different locations reveals its '*distributive* agency,' as well as the process of ordering and territorialization through which its discursive and material aspects come together.[33] But Tambora is not simply multilocational; it is also nonlocational in the sense that it challenges the distinction between different places.[34] After the main eruption on the evening of 10 April, a remarkable and terrifying darkness covered a very large area. It is described with particular precision by the Resident at Grissie:

> I woke in the morning of the 12th, after what seemed to be a very long night, and taking my watch to the lamp found it to be half past eight o'clock, I immediately went out and found a cloud of ash descending; at

> 9 o'clock no day light—the layer of ashes on the terrace before my door at the Kradenan measures one line in thickness; ten A. M.—a faint glimmering of light can now be perceived overhead: half past 10—can distinguish objects 50 yards distant: 11 A. M.—Breakfasted by Candle-light, the birds began to chirrup as at the approach of day: half past 11—can discover the situation of the sun through a thick cloud of ashes; 1 P. M. found the layer of ashes one line and a half thick, and measured in several places with the same results; 3 P. M. the ashes have increased one eighth of a line more; 5 P. M. it is now lighter, but still I can neither read nor write without Candles. (6–7)

Crucial to this passage is the precise and minute demarcation of time and space: the sort of precision that you would find in a sea captain's log. In the absence of the usual temporal markers of day and night—the disruption of normalcy—imperial clock-time becomes a vital way of making sense of what is happening and finding order in disorder.[35] A 'line' is an old British measurement, usually reckoned at one twelfth of an inch. (It was not included in the Weights and Measures Act of 1824.) The measuring of the depth of the layer of ashes can therefore be understood as particularly precise, as the length increases from around 2 mm to 2.5 mm to 2.75 mm. As one would expect of a sensible colonial official, the writer does not panic, but proceeds to make orderly observations of the unusual phenomenon. The colonising subject, therefore, is shown to have the knowledge and technology to comprehend even the strangest and most confusing local conditions, to bring a little light into the darkness. And yet this comprehension only becomes meaningful within an assemblage of other partial accounts mediated by a central intellect.

The extract from Grissie emphasises observations of measurable phenomena rather than aesthetics or emotional affect. To some extent, this approach is true of the document as a whole, but a more excited account is provided by the commander of the Benares:

> The morning of the 12th was extremely dark and lowering [...] At 8 A. M. it was apparent that some extraordinary occurrence had taken place; the face of the heavens to the Southward and Westward had assumed a dark aspect, and it was much darker than before the sun rose; as it came nearer it assumed a dusky red appearance, and spread fast over every part of the heavens; by 10 it was so dark that a ship could hardly be seen a mile distant; by 11 the whole of the heavens were obscured, except a small space near the horizon to the Eastward, the quarter from which the wind came. The ashes now began to fall in showers, and the appearance was

> altogether truly awful and alarming. By noon the light that had remained in the Eastern part of the horizon disappeared, and complete darkness covered the face of day. This continued so profound during the remainder of the day, that I never saw any thing equal to it in the darkest night—it was impossible to see your hand when held up close to your eyes.(13-14)

'Face of the heavens' is a reasonably common phrase in the period, although the particular context in this passage may suggest an allusion to a Biblical passage in which Jesus addresses the Pharisees and Sadducees: 'When it is evening, ye say, It will be fair weather: for the sky is red. And in the morning, It will be foul weather to day: for the sky is red and lowring. O ye hypocrites, ye can discern the face of the sky; but can ye not discern the signs of the times?' (Matthew 16.3). The commander observes a 'red and lowring' dawn, which does indeed presage the foulest of weather. And 'the dark aspect' suggests something threatening and sinister. Unlike the apparently emotionless response of the Resident at Grissie, this witness relays his sense of terror. The emphasis on the 'awful and alarming' conditions, along with the possible Biblical allusion, is perhaps the closest that the narrative gets to any sort of apocalypticism.[36] Again, the epistemological stability provided by imperial clock-time becomes of vital importance in an environment that can no longer be comprehended or even seen.

By bringing together accounts such as these, the narrative attempts to overcome local differences. Its synchronic approach also seeks to overcome the central problem of imperial management in the early nineteenth century: the mediated flow of information. Until the invention of the telegraph, the distance between the centre and the peripheries was not only spatial but temporal. This was a particular problem for colonial policy in the East Indies because it was largely dependent on events happening within Europe. In Raffles's document, the speed of the eruption's effects provides a sort of illusion of contiguity so that it seems simultaneously present in 'every quarter' (1). It becomes a metaphor for what imperialism can only dream of being: an assemblage that transcends the local and is simultaneously present everywhere. Writing in 1834, Thomas De Quincey vividly articulated this fantasy:

> The national voice was lost in the distance, and could not collect itself through the time and space rapidly enough to connect itself immediately with the evanescent measure of the moment. But as the system of

> intercourse is gradually expanding, these bars of space and time are in the same degree contracting, until finally we may expect them altogether to vanish: and then the whole empire, in every part, will react upon the whole through the central forces, with the power, life, and effects of immediate conference amongst parties brought face to face. Then first will be seen a political system truly *organic* – *i.e.* in which each acts upon all, and all react upon each: and a new earth will arise from the indirect agency of this merely physical revolution.[37]

In this remarkable anticipation of globalisation, De Quincey imagines a political system that is also an organic phenomenon: a global body politic in which agency is perfectly distributed. The result is an assemblage of different parts that are instantaneously connected and reactive. From the perspective of Raffles's narrative, Tambora, too, is a phenomenon that is simultaneously present across the archipelago:

> On the night of the 11th the explosions [the Resident of Bima] represents to have been most terrific, and compares them to the report of a heavy mortar close to his ear. (16–17)

> The sound appeared to be so close that in each district it seemed near at hand; it was attributed to an Eruption from the Marapi, the Gunung Klotot or the Gunung Bromo. (4)

These are not simply misunderstandings, but reflect the capacity of the Tambora assemblage to transcend locality: to communicate and impact on a speed and scale far beyond the capacity of nineteenth-century imperialism. The narrative seeks not only to represent this nonlocality but to achieve it, by taking what is apparently a disinterested overview of the different localised accounts.

To maintain the illusion of disinterestedness, it is not until nearly halfway through the narrative that we get any sense of the human cost of the eruption. The text starts on Java, the centre of colonial government, and then moves towards Sumbawa, where the effects were much worse. The narrative's emphasis in its early pages is therefore focused on the sublime natural phenomena experienced across several islands rather than the more local effects on Tambora. The terrible human disaster is not even mentioned in the preamble. Instead, the narrative builds up gradually, creating some suspense as the reader gets closer to the centre of the catastrophe. Six pages into the eighteen-page document, there is a rather sanguine reference to the effects on Javanese agriculture, although the damage to crops and livestock in some areas is acknowledged. After eleven pages, an account by 'the Noquedah of a Malay Prow' who reports landing on Sumbawa and finding 'many of the Inhabitants dead from famine' (11), but this is merely a passing phrase that is easily lost in the grand theatre of natural phenomena that surround it. It is only in the last four pages of the narrative, in Lieutenant Phillips's account from Sumbawa, that readers are given a fuller sense of the eruption's more direct effects on the local population. Phillips was apparently 'well versed in the Malayan language,' and his is the only colonial account to include anything like a proper engagement with an indigenous perspective on the eruption.[38] Each paragraph of his account is in inverted commas, which is clearly meant to suggest that his words are reproduced verbatim. Sent by Raffles to manage the delivery of 'a supply of rice to their relief' and to discover the 'local effects,' Phillips notes that

> The extreme misery to which the inhabitants have been reduced is shocking to behold—there were still on the road side the remains of several corpses and the marks of where many others had been interred—the Villages almost entirely deserted—and the houses fallen down—the surviving inhabitants having dispersed in search of food. (21–22)

In a narrative so lacking in emotion, 'extreme' and 'shocking' have a powerful charge. This sense of total societal collapse continues as Phillips notes that 'a violent Diarrhoea [...] has carried off a great number of people' (22).[39] The severity of the famine is further emphasised by the report of the Rajah of Saugar (the kingdom next to Tambora) that 'one of his own daughters died from hunger.' The Rajah's voice is in the third person and is so far removed from the narrative itself, mediated through Phillips (and through Assey and Raffles), that there is little sense of him as a character to be sympathised with. Phillips notes that the Rajah 'was himself a spectator of the late Eruption' and therefore that his account 'is perhaps more to be depended upon than any other I can possibly obtain' (23). The three paragraphs that follow give the only known eyewitness account of the initial eruption itself. For two of them, which focus on description, we might assume that we are getting a sort of composite of Phillips's and the Rajah's voices. The return to Phillips's colonial perspective is apparent towards the end of the third paragraph, which notes that 'the Natives are apprehensive of another Eruption during the ensuing rainy season' (24).

The Rajah's account is the most dramatic section of the narrative and gives a sense of the extreme power of an elemental assemblage beyond human control:

> The fire and columns of flame continued to rage with unabated fury [...]. Stones at this time fell very thick at Saugar—some of them as large as two fists, but generally not larger than walnuts; between 9 and 10 P. M ashes began to fall and soon after a violent whirlwind ensued, which blew down nearly every house in the village of Saugar, carrying the tops and light parts away with it. In the part of Saugur adjoining Tomboro, its effects were much more violent, tearing up by the roots the highest trees, and carrying them into the air together with men, houses, cattle and whatever else came within its influence (this will account for the large number of floating trees seen at sea). The sea rose nearly 12 feet higher than it had ever been known to be before, and completely spoiled the only small spots of rice lands in Saugur—sweeping away houses and every thing within its reach. (23–24)

The passage emphasises human vulnerability to elemental forces, moving through fire, earth, air, and water. The idea of the fire 'raging' and in 'fury' is closer to anthropomorphism than the narrative usually gets and suggests an agency behind this conflagration. It might be linked to the more celebratory personification earlier in the document where it is reported that the eruption is caused by the deity 'Nyai Loroh Kidul' marking the marriage of one of her children. As in other parts of the narrative, the description of abnormal phenomena is an attempt to exert representational control through the demarcation of time and other measurements (the size of the stones and the height of the sea), as well as grammatical and syntactical ordering, but here it can only seem futile in the context of elemental violence and mortal danger. The destruction of human dwellings and means of sustenance, often so important to representations of catastrophe, is emphasised when the narrative shifts back to Phillips's overview, as he notes that only one of the villages in Tomboro is remaining and that 'in Precate, no vestige of a house is left' (24). He estimates there were 'certainly not fewer than 12,000 individuals in Tomboro and Precate at the time of the Eruption' (24–25); the implication is that the majority of them have perished. However, this suggestion is not developed and is in any case a vast underestimate of the catastrophe's local death toll. Phillips goes on to state that a 'high point' near the eruption avoided the complete destruction of 'trees and herbage' along the 'North and West sides of the Peninsula' and that three people were saved from there on the night of the eruption, although he has not been able to find them (25). This 'search' for the three survivors inevitably exists in an ironic contrast with the 12,000 people mentioned in the previous paragraph and who Phillips believes have been destroyed.

After depicting such devastation, the narrative attempts to end on a hopeful note. In response to the report from a messenger to Sumbawa that 'an immense number of people have been starved,' Phillips notes, in the document's final sentence, that 'the distress has, however, I trust, been alleviated by this time, as the brig with 63 coyangs of Rice from Java arrived there the day he was leaving it' (25). As Wood points out, this amount of aid would only have lasted a few days, although to be fair on Raffles and Phillips they probably had little idea of the scale of the devastation.[40] (The awkward parenthetical caveats 'however, I trust' may suggest Phillips's lack of certainty on this score.) Earlier in the document, the metanarrator notes that the Lieutenant Governor had 'dispatched a supply of rice' for the inhabitants' relief, under Phillips's command (11). The narrative ends, therefore, with a reminder of the supervening intelligence and care of the British administration, which observes unusual phenomena disinterestedly and dispassionately, but also exercises its duty to the indigenous peoples within its sphere of influence. The first publication of the Narrative within the journal of a Society dedicated to producing knowledge that would support colonial development is therefore entirely fitting, and suggests that it does much more than simply provide a neutral account of the eruption, despite the uses to which it has generally been put by scholars. Its heteroglossic form, rather, marks it out as part of the discursive-material Tambora assemblage: an imbrication of living beings, cultural constructions, energy flows, and objects. That is, the narrative is not merely a commentary on a political-ecological catastrophe, but an intervention within it.

Tambora and *The History of Java*

Given the relative obscurity of the ***Batavian Transactions***, British readers are most likely to have encountered the Tambora narrative in volume one of Raffles's *History of Java* (1817), where an edited version appears as an extended footnote to a discussion of mineralogy and volcanism in the first chapter (which focuses on Java's geography, ecology, and climate).[41] As Wood notes, 'the implicit political purpose' of the *History* 'was to argue the case for continued British administration in Java.'[42] Raffles envisages that, following British land reforms, the island would become 'the metropolis, the granary, and the centre of civilisation to the vast regions between the coast of China and the Bay of Bengal.'[43] He emphasises Java's economic potential, the good qualities of its indigenous population and, importantly, its generally excellent climate, noting that the 'general inference' drawn from the experience of the British occupation was that 'with the exception of the town of Batavia, and some parts of the northern coast, the island of Java stands on a level, in point of salubrity, with the healthiest parts of British India, or of any tropical country in the world.'[44] For Raffles, the island's lack of productivity was due to the colonial mismanagement of the Dutch, rather than any intrinsic problems, and its varied climate and rich soil gave it great agricultural potential for cash crops: 'the productions of every region in the world may find a congenial spot somewhere in the island.'[45] Java's volcanism was a double-edged sword. Raffles acknowledges the 'violent convulsions' that it and other nearby islands 'have so often suffered,' but also notes that the remarkable depth and richness of its soil was 'probably owing to the exclusively volcanic constitution of the country.'[46] Wood notes that the placing of the account of the eruption within a footnote to the *History* may suggest Raffles's uneasiness with an account of unpredictable volcanic activity in the region that potentially worked against his emphasis on Java's potential as a productive British colony:

> To have explored in detail the regional impact of ecological destruction, crop failure, famine, disease, death, homelessness, and enslavement within the British colonial domain that were a direct consequence of Tambora would have been to concede the vulnerability of this tropico-georgic paradise to volcanic disaster, and of the Southeast Asian economy to drastic short-term climate change.[47]

However, Wood's suggestion that Raffles seeks to 'wilfully deny' Tambora's impact is exaggerated.[48] The footnote takes up such a large proportion of the three pages that it covers that it is actually rather prominent: it fills around half the first page, and then almost the entirety of the second and third pages. Given that he was writing a history of Java, rather than of the archipelago in general, Raffles did not need to include the Tambora narrative. And a footnote is actually the natural place for the account, as the book itself is not a history in the narrative sense, but a geographical, economic, and anthropological overview of Java. Raffles does not seek to downplay the force or magnitude of the eruption when introducing the account or in his editorial changes, and the fact that he has relatively little to say about its human impacts may be as much due to ignorance as contrivance. Like the original narrative, the footnote is an imperialistic reterritorialization of the Tambora assemblage: an attempt to bring a heterogeneous set of experiences, some personal and some cultural-mythical, into a unified, fact-making whole.

The opening of the footnote manifests a sublime hyperbole that seeks to reassure readers about the capacity of the educated imperialist to make sense of powerful elemental forces, but also reflects the resistance of those forces to a centralising epistemology:

> In order to give the reader some idea of the tremendous violence with which nature sometimes distinguishes the operations of the volcano in these regions and enable him to form some conjecture, from the occurrences of recent experience, of the effects they may have produced in past ages, a short account of the extraordinary and widespread phenomena that accompanied the eruption of the Tomboro mountain, in the island of Sumbawa, in April 1815, may not be uninteresting. Almost everyone is acquainted with the intermitting convulsions of Etna and Vesuvius, as they appear in the descriptions of the poet and the authentic accounts of the naturalist, but the most extraordinary of them can bear no comparison, in point of duration or force, with that of Tomboro. This eruption extended perceptible evidences of its existence over the whole of the Molucca Islands, over Java, a considerable portion of Celebes, Sumatra, and Borneo, to a circumference of a thousand statute miles from its centre, by tremulous motions, and the report of explosions; while within the range of its more immediate activity, embracing a space of three hundred miles around it, it produced the most astonishing effects, and excited the most alarming apprehensions. On Java, at the distance of three hundred miles, it seemed to be awfully present. The sky was overcast at noon-day with clouds of ashes, the sun was envelloped in an atmosphere, whose "palpable" density he was unable to penetrate; showers of ashes covered the houses, the streets, and the fields, to the depth of several inches; and amid this darkness explosions were heard at intervals, like the report of artillery or the noise of distant thunder. So fully did the resemblance of the noises to the report of cannon impress the minds of some officers, that from an apprehension of pirates on the coast vessels were dispatched to afford relief. Superstition, on the other hand, on the minds of the natives,

> was busily at work, and attributed the reports to an artillery of a different description to that of pirates. All conceived that the effects experienced might be caused by eruptions of some of the numerous volcanos on the island; but no one could have conjectured that the showers of ashes which darkened the air, and covered the ground of the eastern districts of Java, could have proceeded from a mountain in Sumbawa, at the distance of several hundred miles. Conceiving that it might be interesting and curious to preserve an authentic and detailed account of the information that could be gained of this wonderful phenomenon, while the event was still recent and full remembered, I directed a circular to the different residents, requiring them to transmit to the Government a statement of the facts and circumstances connected to it, which occurred within their own knowledge. From their replies, the narrative drawn up by Mr. Assey, and printed in the ninth volume of the Batavian Transactions, was collected; the following is an extract of that paper.[49]

The first two sentences move interestingly between what might be termed 'continuous' and 'catastrophic' readings of Tambora. On the one hand, Raffles encourages the reader to place the recent eruption within the context of the history of previous volcanic activity, both within the region and within Europe. This tempers its disastrous nature and suggests the power of a European elite to make sense of it through historical analogy and by placing it within a familiar discourse of volcanism (as well as, more broadly, within a scholarly survey of the region). Thus, the audience is assumed to be familiar with Virgil's description of the eruption of Etna in the *Aeneid*, and Pliny the Younger's letters describing the eruption of Vesuvius. (It is possible that 'the naturalist' also refers to the celebrated eighteenth-century vulcanologist Sir William Hamilton, author of *Observations on Mount Vesuvius, Mount Etna, and Other Volcanos* (1772)). However, on the other hand, the appeal to familiar knowledge is undercut by the assertion that Tambora is 'extraordinary' and that its power and scale dwarfs that of the most notorious European volcanos. The superlatives continue with the claims that the eruption produced the 'most astonishing effects' and 'most alarming apprehensions.' Its nonlocality is emphasised by the claim that it is 'awfully present' in Java, three hundred miles away. Its awfulness (that is, its capacity to create terror) is not only due to the force of the eruption, but its ability to penetrate space, apparently instantaneously.

Raffles goes on to describe the disruption of normality that is the key trope of catastrophe narratives. Not only is the sky dark at noon, but this is a peculiarly 'palpable' darkness: one that that you can almost grab hold of. This is consonant with the accounts of light loss that we find in the original narrative but which are not reprinted in the *History*. By putting 'palpable' in quotation marks, Raffles makes an overt allusion to the scene in Book II of *Paradise Lost* in which the devils in Pandaemonium are debating the best course of action. Beezlebub, Satan's mouthpiece, suggests taking vengeance on God by corrupting the world that he has created,

> But first whom shall we send
> In search of this new world, whom shall we find
> Sufficient? who shall tempt with wand'ring feet
> The dark unbottom'd infinite Abyss
> And through the palpable obscure find out
> His uncouth way, or spread his airy flight
> Upborne with indefatigable wings
> Over the vast abrupt, ere he arrive
> The happy Isle.[50]

As previously noted, it was common in the period to refer to Milton when representing the sublime, but the specifics of Raffles's allusion are worth considering. 'The happy Isle' is of course Earth, the 'new world' created by God with his favoured creatures in it. Penetrating the dark abyss—'the palpable obscure'—between Pandaemonium and Earth will lead Satan into Eden. Similarly, the reader has to see through the dark veil of Tambora—an epistemological as well as a sensory phenomenon—if they are to arrive at an understanding of the potential of the 'happy Isle' of Java. The volcano threatens the abyssal and destructive, but it can also be transcended and controlled by a 'sufficient' and 'indefatigable' hero.

Raffles is interestingly ambivalent about the purpose of including the account of Tambora in the *History*. The penultimate sentence suggests the kind of casual interest in recording 'wonderful phenomena' evident in many publications by gentleman naturalists in the period. But the second half of the sentence moves from amateurish enthusiasm to the control of information demanded by professional empire-builders: Raffles 'directed' and 'required' that knowledge be produced. Two other things are notable about this preamble. First, the eruption is predominantly understood aesthetically and scientifically: Raffles is concerned with it as a sublime natural wonder. There is no mention of the human consequences, although more detail is given on these in the eyewitness account. The other thing is that Raffles steadfastly refuses to draw any religious conclusions from the catastrophe. He does not at any point mention the workings of Providence that were so often a feature of discussions of catastrophic phenomena in the eighteenth and early nineteenth centuries. The document is not concerned with causes, telluric or heavenly, but with experiential phenomena. To return to the distinctions made by Jasanoff and Rigby, it eschews 'meaning-making' moral-mythic interpretations in favour of a more scientific fact-making mode.

Following his preamble, Raffles includes most of the document's metanarrative, presented as composed by Charles Assey, but cuts the extracts from the letters from Grissie, Sumanap, Baniowangie, Fort Marlboro and Bima, as well as the reports from the 'Malay Prow' and the 'Cruizer Benares.' The only direct eyewitness account included is Phillips's account from Sumbawa itself (which contains within it the Rajah of Saugar's eyewitness account of the eruption). The extract from Phillips's letter is included more or less as it is in the original narrative, with some changes to the paragraphing, and with the last 140 words or so cut. The footnote narrative therefore differs from the original by placing less emphasis on the sublime effects of the eruption for those remote from it, and more on its direct effects on the inhabitants of Sumbawa. Phillips's narrative of famine and destruction is much more strongly to the fore. Raffles includes Phillips's devastating statement that 'From the most particular inquiries I have been able to make, there were certainly not fewer than twelve thousand individuals in Tambora and Pekate at the time of the eruption.' Crucially, unlike the original narrative, the footnote includes the second clause of this sentence: 'of whom only five or six survive.'[51] Wood's suggestion that Raffles 'makes no specific acknowledgement of the devastating human impact of the Tambora eruption in his *History of Java*' may be technically correct, but it gives a misleading impression of the acknowledgement that he does make by including Phillips's narrative.[52] Raffles's footnote, whatever the intention of the *History* as a whole, reflects the vulnerability of human populations in an area of unpredictable volcanic activity. Furthermore, by cutting the last few sentences, the footnote version ends with a vision of destruction, rather than the more uplifting vision of the 'alleviation' of the inhabitants' distress through the provision of aid.[53] Wood's suggestion, therefore, that 'in the *History*, Raffles appends the narrative of the Tambora eruption as a "curiosity," not the epochal human tragedy it truly was,' does not seem quite fair.[54] The tragedy of Tambora is registered, although in a subdued way, and it seems likely Raffles himself did not have a sense of the full scale of the eruption's effects.

Raffles's *Memoir* and the Politics of Catastrophe

It should be apparent by now that the Tambora narrative encodes a set of cultural-political assumptions that are inflected by the context of its publication. This inflection is also apparent in my final case study, which focuses on the 1830 reprint of the narrative in the eighth chapter of the rather hagiographic memoir—largely a collection of Raffles's letters—produced by his widow. An analysis of this textual manifestation brings out the relationship between the disaster and the Anglo-French struggle for global imperial ascendancy. Bar one minor cut and some typographical changes, the Narrative is repeated verbatim, and with only the following editorial commentary: 'Mr. Raffles gives the following account of the eruption of the Tomboro Mountain, in the Island of Sambawa, which took place at this time (the 11th and 12th of April, 1815,) one of the most violent and extraordinary of such explosions yet known.'[55] The implication is that the narrative stands as a sign of Raffles's wide intellectual interests, his efficacy as a colonial administrator, and his care for the indigenous population, as represented by the aid that he provides to the victims of the eruption. Later in the chapter, the editor describes 'the deep interest which Mr. Raffles took in the happiness of the Javanese' (261). What gives the narrative a different resonance in the context of the *Memoir*, as compared to the *Batavian Transactions* or the *History of Java*, is the fact it is followed, without any intervening commentary, by extracts from three private letters that show Raffles's concern with the future of the British empire in Indonesia in relation to global political events and particularly Napoleon's resurgence. Addressing the Orientalist scholar William Marsden (who had once served as First Secretary to the Admiralty) on 6 August 1815, he writes that 'the wonderful events in Europe still leave some hope that Java may remain permanently English' (250) and signs off by asking that Marsden 'advocate[s] the cause of Java, if there is a possibility of its remaining under British protection' (251). This is immediately followed by extracts from a letter to W.B. Ramsay, written the previous day, which gives a clearer sense of exactly what opportunity Raffles saw in the 'the wonderful and extraordinary change in the politics of Europe, [that] by the re-appearance of

Buonaparte, has, with all its horrors, shed one consoling ray on this sacred Isle; and Java may be yet permanently English' (251).The escape of Napoleon from Elba is a horrific catastrophe that—by voiding the Treaty of Paris—nonetheless creates opportunities for the British empire and the Javanese people who, for Raffles, would be much better off with the British than the Dutch (and their French imperial masters). Under such trying circumstances, Raffles hopes, the East India Company and the British government might see the value of having a permanent colonial presence in the area, rather than handing Java to their enemies. The metaphor of the 'consoling ray' seems to suggest that even though Napoleon's resurgence has brought darkness to Europe, it offers the possibility of enlightenment to Java. Raffles's fervent belief in Java's potential as part of the British Empire is apparent in his description of it as a 'sacred Isle': a probable allusion to the description of England as 'this sceptred isle [...] this other Eden' in *Richard II* (as well as, perhaps, the 'happy Isle' of Earth in *Paradise Lost*).[56]

Raffles notes to Ramsay that he has also 'addressed Lord Buckinghamshire, both officially and privately' (251). Robert Hobart, Lord Buckinghamshire, was President of the Board of Control in the Liverpool administration and Raffles's superior. (The Board oversaw the East India Company and Britain's affairs in India.) The *Memoir* reproduces a long letter from Raffles to Hobart that begins by noting that Napoleon's reappearance and the Declaration of the Congress of Vienna 'completely deranges every plan connected with the pacific arrangements about to have been concluded' (252). The letter is an attempt to assert the economic and strategic value of Java to Britain now that real peace seems to be a long way off: 'England must doubtless hold what now she has, and gain what more she can' (252). Raffles argues that his reforms have made the island a potentially lucrative possession and he suggests ways of developing the colonial infrastructure, including moving the seat of government from the unhealthy Batavia to Samarang. The letter also emphasises that the 'Malays' in general 'are very far from being savages' (256) and that the Javanese would very willingly consider themselves British subjects:

> There are no people more capable of appreciating a benefit or forming an attachment; they have no prejudices of moment, or which stand in the way of civilization, and they will readily assimilate with Europeans as far as Europeans are inclined to assimilate with them; they will identify themselves as British subjects whenever they feel the benefit of British principles and the permanency of British power. (259)

For Raffles, the indigenous inhabitants of Java are perfect colonial subjects, without their own opinions or 'prejudices,' and more than willing to attach themselves to a suitably paternalistic power. The letter is entirely in line with the arguments that he would go on to make in the *History of Java*, that the Dutch had failed in their duties as enlightened colonists, and that Britain had put this right:

> When I assumed the delegated administration of Java, humanity dictated consideration and respect for its European inhabitants, but policy required and justice demanded protection to the native population. This protection has been given, and is felt in the cottage of the peasant; he acknowledges, and is proud of the benefit, and in *one word*, the Javanese are decidedly English; give but the *other word*, and Great Britain produces not more faithful adherents to the crown than Java may afford. (260)

The idea that the Javanese have been transformed into grateful subjects with little difficulty is emphasised by Raffles's allusion to the paternalistic ideal of protecting the homes of the rural labouring classes ('the cottage of the peasant'). He claims that his enlightened administration has effectively made them 'English' (the '*one word*'); if Hobart is willing to reconsider and affirm the long-term future of Java as a British colony (giving the '*other word*') then he can be assured of their loyalty and gratitude.

The irony of these letters is that by the time Raffles had heard the news of Napoleon's reappearance, the former Emperor had already been defeated at Waterloo.[57] They reveal starkly the haphazard nature of imperialism in the period, when there was often a strong mismatch between the knowledge and objectives held by the centre and those of the peripheries. Imperialism in the early nineteenth century was a misfiring assemblage in which agency was often inefficiently distributed. Proponents of empire sought to transcend the local but were continually mired in it due to the sluggishness of communications. The ability of an assemblage like the Tambora eruption to be present everywhere simultaneously clearly fascinated Raffles. As Lieutenant-Governor of Java, he was able to commission a report that represented a significant moment of cultural-political territorialization: an attempt to control the Tambora assemblage and mimic its apparent synchronicity. The importance of the narrative lies in much more than what it tells us about the eruption's physical processes or even its effects on the local population. Its compilation of various experiences and interpretations reveals some of the complexity of human-nonhuman entanglements that create an environmental catastrophe and the modern shift from moral-mythic interpretations of such occurrences to more scientific and supposedly disinterested accounts. My analysis of its publication in three different print contexts reveals not only the manifold imbrications of world and text, but the changing political valencies of those imbrications. In the following chapter, the global scope of the material-discursive Tambora assemblage is further addressed through an analysis of the Diodati Circle's 1816 writings. These texts certainly reflect the cultural politics of catastrophe and, like Raffles's account, reveal an elite perspective on a situation from which their authors were largely protected. They also respond to the climate crisis with some remarkable speculations about the extinction of the entire human species.

Notes

1. [Charles Assey], 'Narrative of the Effects of the Eruption from the Tomboro Mountain, in the Island of Sumbawa, on the 11th and 12th of April 1815,' *Transactions of the Batavian Society, of Arts and Sciences* 8 (1816): 3–4. Further references are in the text. The Tambora eruption and its aftermath were not widely reported. An instructive comparison can be made with the publicity generated by the much smaller Krakatoa eruption of 1883 due to the improvements in global communications in the intervening period, most notably the invention of the telegraph and the laying down of oceanic cables. Nonetheless, fragments of information about the eruption and its effects appear in English-language periodicals, taken from a number of sources, including Raffles's narrative.
2. Daniel Defoe's *The Storm* (1704) may have offered something of a prototype for the Tambora narrative. Defoe's innovative work is similarly composite and concerned with harnessing a range of eyewitness accounts in order to construct a wide-ranging and authoritative account of a catastrophic phenomenon. See Daniel Defoe, *The Storm*, ed. Richard Hamblyn (London: Penguin, 2005). I am grateful to Tess Somervell for suggesting this comparison.
3. Kate Rigby, *Dancing with Disaster* (Charlottesville and London: University of Virginia Press, 2015), 3.
4. I have taken this distinction from Sheila Jasanoff, 'A New Climate for Society,' *Theory, Culture & Society* 27 (2010): 233–253. I am grateful to Adeline Johns-Putra for drawing Jasanoff's work to my attention.
5. Jane Bennett, *Vibrant Matter: A Political Ecology of Things* (Durham and London: Duke University Press, 2010), 23.

6. Bennett, *Vibrant Matter*, 24.
7. Bennett, *Vibrant Matter*, 34.
8. Serenella Iovino and Serpil Oppermann, 'Introduction,' in *Material Ecocriticism*, eds. Serenella Iovino and Serpil Oppermann (Bloomington and Indianapolis: Indiana University Press, 2014), 1–17 (3).
9. Timothy Morton, *Hyperobjects* (Minneapolis and London: University of Minnesota Press, 2013), 1.
10. Iovino and Oppermann, 'Introduction,' 9. Karen Barad's work has proved particularly influential to this approach: see *Meeting the Universe Halfway: Quantum Physics and the Entanglement of Matter and Meaning* (Durham and London: Duke University Press, 2007), especially Chap. 4.
11. I am grateful to Adeline Johns-Putra for drawing this to my attention.
12. Gilles Deleuze and Felix Guattari, *A Thousand Plateaus: Capitalism and Schizophrenia*, trans. Brian Massumi (London: Bloomsbury, 2016), 586.
13. Deleuze and Guattari, *Thousand Plateaus*, 587.
14. There is also a fourth version of the narrative, which I read from a bound volume of the *Transactions* held by the Royal Society. Apart from different pagination and a few typographical errors, it is identical to the original published version.
15. Iovino and Oppermann, 'Introduction,' 8.
16. For reasons of space, focus, and my own expertise, this chapter does not discuss indigenous perspectives on Tambora beyond those featured in the narrative. Wood (24–26) reports on present-day stories told by Sumbawans about the wealthy kingdom of Tambora before the *zaman hujan au* (time of the ash rain). Bernice de Jong Boers quotes from the indigenous poem *Syair Kerajaan Bima*, composed around 1830, which describes the eruption as 'the wrath of God Almighty / At the deed of the king of Tambora / In murdering a worthy pilgrim' (37).
17. The journal's Dutch title is *Verhandelingen van hat Bataaviaasch Genootschap, der Kunsten en Wetenschappen.*
18. Assey presumably left the administration when the island was returned to Dutch control in 1816. The same year, he published a *Review of the Administration, Value, and State of the Colony of Java.*
19. See, for example, de Jong Boers, 37, and Wood, 239.
20. The almanac is available online at https://sites.google.com/site/sumatraswestkust/java-almanac-1, accessed 30 May 2016.
21. Raffles heard that he was likely to be replaced in September 1815 and received news of the final defeat of Napoleon and the return of Java to the Dutch in early November 1815. For his last months in Java, see Victoria Glendinning, *Raffles and the Golden Opportunity* (London: Profile, 2012), Chap. 7.

22. The Klingamans make no reference to Ross or Assey, and seem to assume that Raffles is the sole author of the document (1–4).
23. Despite its heteroglossia, it only refers to one person by name (Lieutenant Owen Phillips).
24. Some of the information in this paragraph is taken from Eloise Smith Van Niel, *The Batavian Society and Scholarship in the Dutch East Indies, 1778–1850* (MA Thesis, University of Hawaii, May 1980), Chap. 3.
25. 'New Regulations for the Batavian Society, of Arts and Sciences' are printed in the seventh volume (1814) of the *Transactions* (xiv-xvii).
26. 'New Regulations', xiv.
27. Thomas Stamford Raffles, 'A Discourse Delivered at a Meeting of the Society of Arts and Sciences in Batavia, on the Twenty-fourth day of April 1813, being the Anniversary of the Institution,' *Transactions of the Batavian Society, of Arts and Sciences* 7 (1814): 26–27.
28. On two occasions, the corporate voice is replaced by a rather jarring first-person reference to the shadowy metanarrator: 'no description of mine' (6); 'I shall conclude this concise narrative' (21).
29. John Milton, *Paradise Lost*, ed. Christopher Ricks (London: Penguin, 1968), 22 (I.593–599).
30. This passage is also alluded to in Gilbert White's description of the fog caused by the Laki volcanic eruption in 1783: see *The Natural History of Selborne*, ed. Richard Mabey (London: Penguin, 1987), 265–266.
31. For the connected rhetorics of weather and warfare in the Romantic period, see Mary A. Favret, *War at a Distance: Romanticism and the Making of Modern Wartime* (Princeton and Oxford: Princeton University Press, 2010), Chap. 3.
32. For a suggestive account of mists, mediation, and allusion in Romantic literature, see Tess Somervell, 'Mediating Vision: Wordsworth's Allusions to Thomson's *Seasons* in *The Prelude*', *Romanticism* 22 (2016): 48–60 (51–55).
33. Bennett, *Vibrant Matter*, 21.
34. I have taken the term 'nonlocality' from Morton but use it in a slightly different way from him, in part because he is keen to distinguish hyper-objects from assemblages (*Hyperobjects*, 2).
35. For the imposition of western time-keeping on colonial cultures, see Giordano Nanni, *The Colonisation of Time: Ritual, Routine, and Resistance in the British Empire* (Manchester and New York: Manchester University Press, 2012).
36. Some more religious interpretations of Tambora can be found in periodical accounts; for example, 'Philomath,' a correspondent published in the *Methodist Magazine*, prefaces a reprinted newspaper article on the

eruption with the rhetorical question, 'O when will the inhabitants of the world learn righteousness, seeing the judgements of God are abroad in the earth!': see 'The Works of God Displayed: An Account of a Volcanic Eruption,' *Methodist Magazine* December 1816: 910–911 (910).

37. Thomas De Quincey, 'Travelling in England Thirty Years Ago,' in *The Works of Thomas De Quincey*, ed. by Grevel Lindop et al., 21 vols (London: Pickering and Chatto, 2000–2003), X, ed. by Alina Clej, 96.
38. John Crawfurd, *History of the Indian Archipelago*, 3 vols (Edinburgh: Constable, 1820), II, 124.
39. Phillips notes that the 'Natives' suppose this to be caused by drinking water impregnated with ash and that horses had suffered a similar fate. For the human tragedy of Tambora, see Oppenheimer, *Eruptions*, 308–312.
40. Wood, *Tambora*, 31. A coyang is 3300 lb, which makes a total of 207,900 lbs of rice. Given around a pound of rice per person per day, this could plausibly feed 10,000 people for twenty days. (The United Nations World Food programme provides, on average, a little under a pound of rice per person per day.)
41. They might also have found fragmentary news of Tambora in periodicals and in reviews of the *History*. I have found a total of nine original ones: in the *British Review*, the *Eclectic Review, Edinburgh Monthly Review*, the *Edinburgh Review*, the *Literary Gazette*, the *Literary Panorama*, the *Monthly Magazine*, the *Monthly Review*, the *Quarterly Review*. Three of these mention Tambora. The *British Review* quotes the opening and closing paragraph from the footnote, the *Eclectic Review* quotes part of Raffles's preamble, and the *Monthly Review* transcribes the whole of the footnote except for the opening paragraph.
42. Gillen D'Arcy Wood, 'The Volcano Lover: Climate, Colonialism, and the Slave Trade in Raffles's *History of Java* (1817)', *Journal for Early Modern Cultural Studies* 8 (2008): 33–53 (36).
43. Thomas Stamford Raffles, *The History of Java*, 2 vols (Kuala Lumpur: Oxford University Press, 1965), I, 161. Reprint of 1817 edition.
44. Raffles, *History*, I, 31.
45. Raffles, *History*, I, 34.
46. Raffles, *History*, I, 24, 29.
47. Wood, *Tambora*, 31. An earlier lengthy footnote in the same chapter includes an account of three Javan volcanoes by the naturalist Robert Horsfield, which describes the eruption of Papandayan in 1772 that reportedly killed nearly 3000 people: Raffles, *History*, I, 15.
48. Wood, *Tambora*, 31.
49. Raffles, *History*, I, 25–26.
50. Milton, *Paradise Lost*, 39 (II.402–410).

51. Raffles, *History*, I, 28.
52. Wood, 'The Volcano Lover,' 42.
53. I suspect that Raffles cut this reference so that the footnote would not appear self-serving in what was meant to be a disinterested, scholarly publication.
54. Wood, 'The Volcano Lover', 49.
55. Sophia Raffles, ed., *Memoir of the Life and Public Services of Sir Thomas Stamford Raffles* (London: John Murray, 1830), 241. Further references to the *Memoir* are in the text.
56. William Shakespeare, *The Life and Death of King Richard the Second*, in *Complete Works*, eds. Jonathan Bate and Eric Rasmussen (London: Palgrave, 2008), 847 (II.i.40–42). England is also described as 'this fortress built by nature for herself / Against infection' (II.i.43–44), which resonates with Raffles's emphasis on Java's general healthiness.
57. *En route* to England in May 1816, Raffles convinced his ship's captain to stop at St Helena so that Raffles could meet with Napoleon. From his own account, he seems to have been rather disappointed with the defeated emperor (*Memoir*, 277).

CHAPTER 3

Geohistory, Epistemology, and Extinction: Byron and the Shelleys in 1816

Abstract This chapter sheds new light on the complexity with which Byron and the Shelleys wrote about environmental catastrophe in 1816. They knew nothing of Tambora, but the bad weather that it largely caused was an important influence on their creativity, in combination with their interest in contemporary natural philosophy and their experience of the sublime landscapes around Geneva. This chapter brings together their 1816 writings to reveal the richness of their reflections on the vulnerability of human communities living with uncontrollable geophysical and climatic forces, the entanglement of humans and nonhuman nature, and the possibility of human extinction. Key texts discussed include Byron's *Manfred*, *The Prisoner of Chillon*, and 'Darkness', Mary Shelley's *Frankenstein*, and Percy Bysshe Shelley's 'Mont Blanc', as well as letters and journals.

Keywords Lord byron · Percy bysshe shelley · Mary shelley
Speculative realism · Extinction · Geology

Introduction

In a letter dated 17 May 1816, Mary Shelley paints an idyllic picture of the holiday that she and Percy Bysshe Shelley were taking at the Hôtel d'Angleterre in Sécheron, just outside Geneva: 'You know that we have just escaped from the gloom of winter and of London; and coming

D. Higgins, *British Romanticism, Climate Change, and the Anthropocene*,
DOI 10.1007/978-3-319-67894-8_3

to this delightful spot during this divine weather, I feel as happy as a new-fledged bird.'[1] By the start of June, however, the weather was very different: 'unfortunately we do not now enjoy those brilliant skies that hailed us on our first arrival to this country. An almost perpetual rain confines us principally to the house; but when the sun bursts forth it is of a splendour and heat unknown in England. The thunder storms that visit us are grander and more terrific than I have ever seen before.'[2] This sublime changeability was the prelude to a two-year period of unusual climatic conditions and famine in Europe and across the world. As I discussed in the introduction, scholars have noted the importance of the global climate crisis created by the Tambora eruption to the 1816 writings of the Shelleys and their friend Lord Byron, particularly 'Darkness,', 'Mont Blanc,' and *Frankenstein*. However, this insight has tended to generate broad contextual readings rather than close analysis. This is true even of the astute discussion of the Diodati Circle and the 'Year without a Summer' in Gillen D'Arcy Wood's study of Tambora.[3] My aim in this chapter is to shed new light on the complexity with which Byron and the Shelleys wrote about environmental catastrophe in 1816. They knew nothing of Tambora, but the bad weather that it largely caused was an important influence on their creativity, in combination with their interest in contemporary natural philosophy and their experience of the sublime landscapes around Geneva. This chapter brings together their 1816 writings to reveal the richness of their reflections on the vulnerability of human communities living with uncontrollable geophysical and climatic forces, the entanglement of humans and nonhuman nature, and the possibility of human extinction.

Table 3.1 provides a detailed chronology for the summer of 1816. Byron left England on 25 April, and travelled slowly through Europe before arriving in the Geneva area on 25 May. The Shelley group, including Mary's stepsister (and Byron's former lover) Claire Clairmont, had left on 3 May but only took ten days to get to Geneva. The two parties met on 27 May and spent a considerable amount of time together until the Shelleys left for England on 29 August. During this period, Byron wrote a number of poems, including 'Darkness' (1816), 'Prometheus' (1816), *The Prisoner of Chillon* (1816), and Canto III of *Childe Harold's Pilgrimage* (1816), and began *Manfred* (1817). Mary began *Frankenstein* (1818), following the famous ghost-story competition between the three authors and Byron's doctor John William Polidori, and drafted much of the novel before returning to England.

Table 3.1 Chronology of the summer of 1816

Date	*Byron*	*MS*	*PBS*
25 April	Leaves Dover for Ostend with Polidori; begins *CHP* III at sea		
3 May		Leave Dover with son William and Claire Clairmont	Leave Dover with son William and Claire Clairmont
4 May	Visits the field of Waterloo; continues *CHP* III		
9 May	*Glenarvon* by Caroline Lamb published		
13 May		Arrive at Sécheron near Geneva	Arrive at Sécheron near Geneva
25 May	Arrives at Sécheron near Geneva		
27 May	Meets the Shelleys near Geneva	Meet with Byron	Meet with Byron
By 1 June		Have moved into Maison Chapuis	Have moved into Maison Chapuis
8 June	Finishes first draft of *CHP* III		
10 June	Moves into Diodati		
14/15 June	Ghost stories	Ghost stories	Ghost stories
16–17 June		Begins writing *Frankenstein*	
18 June	Gives evening recital of *Christabel*		
22–30 June	Boating tour		Boating tour; composes 'Hymn to Intellectual Beauty'
27–29 June	Drafts *Prisoner of Chillon*		
4 July	Completes *CHP* III		
12 July	Makes first of several visits to Germaine de Staël at Coppet		
21–27 July		Begin trip to Chamonix	Begin trip to Chamonix
22 July	Sends 'Monody on the Death of Sheridan' to Douglas Kinnaird	See Mont Blanc	See Mont Blanc; begins journal-letter to Peacock

(continued)

Table 3.1 (continued)

Date	*Byron*	*MS*	*PBS*
23 July			Gives this date for composition of 'Mont Blanc'
24 July		Notes in her journal, 'write my story'—her first written reference to *Frankenstein*	
25 July		See Mer de Glace	See Mer de Glace
27 July		Return from Chamonix	Return from Chamonix
14 August	Matthew "Monk" Lewis visits Diodati		
21 August		Notes in her journal, 'Shelley & I talk about my story'—most likely discussed turning the story into a longer book	
26 August	Hobhouse and Scrope Davies arrive		
29 August	Travels to Chamouni with Polidori, Hobhouse, and Davies	Leave Geneva	Leave Geneva
30 August	Sees Mont Blanc and Mer de Glace		
8 September		Arrive at Portsmouth	Arrive at Portsmouth
By 16 September	Has dismissed Polidori		
17 September	Byron and Hobhouse begin tour of the Alps and Byron starts Alpine Journal		
28 September	Finishes journal		
29 September	Return to Diodati		
5 October	Leave Diodati		

(No known dates for 'Darkness,' 'The Dream,' 'Prometheus,' or *Manfred*)

Percy wrote several lyrics, most importantly 'Hymn to Intellectual Beauty' (1817) and 'Mont Blanc' (1817). This last poem was first published at the end of the Shelleys' *History of a Six Weeks' Tour* (1817), which describes a journey that they had taken through Europe in 1814 and their impression of the area around Geneva in 1816. Although the *History* is the only one of these texts to present itself as a collaborative work, the group worked so closely together during this period that we should understand all of their 1816 writings as involving elements of collaboration. As in Chap. 2, I am concerned with a complex assemblage of texts, for addressing the group's individual writings in isolation from each other is to miss out on much of their richness.[4] I understand them as the collaborative products of a creative community and show that they share a concern with the fragility of human dwelling within a potentially violent universe. In articulating this concern, they draw on a shared language of catastrophe—a kind of textual ecology—that repeats, with different inflections, key images and tropes. There is nothing utopian about this ecology; in fact, it often reveals a breakdown of human structures. The two related forms of community that it tends to show as most resilient are, first, the relationship between the individual subject and the nonhuman objects of the sublime landscape, and, secondly, interactions within an elite group of intellectuals whose capacity simultaneously to comprehend and transcend the 'natural' marks them out from the majority of people. Due to wealth and rank, Byron and the Shelleys were largely protected from the terrible effects that the climate crisis had on the poorer inhabitants of Europe. Like Raffles, therefore, their position within the catastrophic assemblage was one of relative safety. As Dipesh Chakrabarty has famously argued, although the history of climate change is intertwined with the history of inequality, it cannot be reduced to that history because ultimately it threatens the whole of humanity.[5] Byron and the Shelleys speak to the concerns of the Anthropocene as some of the very first modern Western thinkers to contemplate the finitude of the human species detached from any eschatological narrative.

The chapter begins with a discussion of the invocation of 'deep time,' and particularly the global cooling theory of the Comte de Buffon, in the *History of a Six Weeks' Tour* and 'Mont Blanc.' The potential threat that glacial augmentation poses to human 'dwelling' is explored in a number of 1816 texts and is often represented through the image of the destruction of pine forests that stand metonymically for human vulnerability. I focus particularly on the suggestion in *Frankenstein* that the Creature may be better equipped to flourish in the desolate world imagined by Buffon, and on the disturbing resistance of ice to human attempts to write themselves on to the landscape. Byron and the

Shelleys, I argue, understand ice as manifesting agency: as 'a restless activeness, a destructive-creative force-presence.'[6] Ecological precariousness and the relationship between human beings and elemental forces are also important to my discussion of Byron's 'Darkness.' I suggest that Byron is sceptical about the possibility of any form of positive collective response to environmental catastrophe, in part because his poetry is troubled by the idea of human consciousness as part of an embodied community of creatures. The notion that human beings (or perhaps an exceptional subset of them) are profoundly caught between the spiritual and earthly realms is most powerfully articulated in *Manfred*, which seeks to assert the power of the individual imagination against a hostile universe. The chapter ends with an extended analysis of 'Mont Blanc' that draws on recent philosophical work in speculative realism and object-oriented ontology.[7] Particularly useful to my approach are Quentin Meillassoux's critique of 'correlationism' ('the idea according to which we only ever have access to the correlation between thinking and being, and never to either term considered apart from the other'), Raymond Brassier's discussion of extinction as it 'indexes the thought of the absence of thought,' and Timothy Morton's analysis of 'hyperobjects' ('things that are massively distributed in time and space relative to humans').[8] My analysis marries epistemological, political, and ecological approaches to the poem by examining how its concern with perception and absence evokes the difficulty of imagining an earth without humanity.

Deep Time and Global Cooling

Towards the end of the eighteenth century, natural philosophers, aware that the earth was considerably older than six thousand years, began to tell the story of prehuman history. Three thinkers about deep time were particularly influential on Byron and the Shelleys.[9] One is the Comte de Buffon, who in 1749 had 'set out a module of continuous but directionless terrestrial change [...] a kind of dynamic equilibrium' in the first volume of his epic *Histoire Naturelle*.[10] However, in 'Les époques de la nature' (1778) and other later works, he presented a quite different narrative, describing how the planet had undergone a process of gradual cooling since its creation and imagining an icy future in which it would be rendered uninhabitable.[11] Byron was particularly responsive to the work of Buffon's rival and compatriot Georges Cuvier, who eventually concluded from his study of fossils that the earth had experienced several catastrophic geological upheavals and extinction events.[12] Finally, Nigel Leask has identified the influence of the Scottish geologist James Hutton on Percy Shelley.[13] His long and esoteric *Theory of the Earth* (1795) had been made accessible by John Playfair's *Illustrations of the Huttonian Theory of the Earth* (1802), and argued against catastrophists like Cuvier by presenting the earth as a body in dynamic homeostasis designed to ensure its habitability for humans.[14] Romantic-period geology was riven with debate, but what was not in doubt was the capacity of landscapes to change over long and short periods through natural processes. In the eighteenth century, sudden environmental catastrophes like the Lisbon earthquake of 1755 had problematised providential and theodical thinking and—despite the efforts of deists such as Hutton—geological developments were continuing this process.[15] The tendency in much Romantic literature and philosophy to valorise the autonomous, self-willing subject may be understood partly as a response to this challenge to anthropocentrism. As Nigel Clark states, 'the trouble with the newly unfurling temporal spans, along with the previously established immensity of cosmic space, was that it implied whole domains of existence in which no humans were present, vast stretches in which other-than-human objects were left to their own devices.'[16] The 1816 writings of Byron and the Shelleys show a profound attempt to make sense of a universe in which the human species seemed to be in danger of moving from the centre to the margins.

The area around Geneva was a particularly rich environment for thinking about deep time. As Europe's high point, Mont Blanc was fascinating to geologists like Horace-Bénédict de Saussure who in 1787 had been one of the first climbers to ascend the mountain and who used his observations of its strata as the basis for what Martin Rudwick calls 'a vivid narrative of imaginatively witnessed geohistorical events.'[17] By 1816, there was also increasing interest in the geomorphic power of glaciers.[18] In the third letter to Thomas Love Peacock published in the *History of a Six Weeks' Tour*, Percy Shelley suggests that it was likely that the Chamonix glaciers were perpetually increasing in size, rather than waxing and waning (as Saussure had argued): 'these glaciers flow perpetually into the valley, ravaging in their slow but irresistible progress the pastures and the forests which surround them, performing a work of desolation in ages, which a river of lava might accomplish in an hour, but far more irretrievably' (158).[19] The desolating power of ice is a key trope in the group's texts of 1816, and Percy goes on to consider the global implications of this local observation, noting in a much-cited passage that

> It is agreed by all, that the snow on the summit of Mont Blanc and the neighbouring mountains perpetually augments, and that ice, in the form of glaciers, subsists without melting in the valley of Chamouni during its transient and variable summer. If the snow which produces this glacier must augment, and the heat of the valley is no obstacle to the perpetual existence of such masses of ice as have already descended into it, the consequence is obvious; the glaciers must augment and will subsist, at least until they have overflowed this vale. I will not pursue Buffon's sublime but gloomy theory — that this globe which we inhabit will at some future period be changed into a mass of frost by the encroachments of the polar ice, and of that produced on the most elevated points of the earth. (161–162)[20]

An earlier section of the letter addresses the sublime power of this landscape on the receptive individual, emphasising its capacity to produce an 'extatic wonder' (152) in a passage that would be further developed in 'Mont Blanc.' Here, however, this power is not understood only in aesthetic terms, but also in relation to Buffon's ideas about deep time and humanity's future. As part of his argument for the influence of Huttonian geology on 'Mont Blanc,' Leask claims that critics have been wrong in assuming that Percy endorses 'Buffon's theory of a returning ice-age,' which he '*flippantly* attributes to Peacock', and relates the letter to Playfair's critique of Buffon's 'dismal and unphilosophic vision.'[21] However, while Percy does not straightforwardly support Buffon's theory, he clearly finds it worthy of serious consideration based on the empirical observation of glacial augmentation.[22] He considers it a 'sublime' theory, one suspects, for several reasons: its intellectual ambition, its assumption of a vast timescale of geological change, its applicability as a model to planets across the universe, and the dark pleasure that one might take in imagining future destruction from a position of present-day safety. To that extent, the implications of Buffon's theory are still aestheticized, and Percy's pleasure here is very much the pleasure of being contradicted. When the poet himself contemplated future climate catastrophe in other texts, it tended to be in the utopian terms of an eternal spring or summer that signifies political liberation, as can be seen in *Queen Mab* (1813) or *Prometheus Unbound* (1820).[23] His interest in Buffon's more 'gloomy' apprehension is the product of a particular place, a particular set of environmental conditions, and a particular creative community, and (as I will show) actually ignores Buffon's rather optimistic view about the human capacity to change the climate.

In fact, Percy goes on to make some pessimistic suggestions about how global cooling might affect the human species:

> Do you, who assert the supremacy of Ahriman, imagine him throned among these desolating snows, among these palaces of death and frost, so sculptured in this their terrible magnificence by the adamantine hand of necessity, and that he casts around him, as the first essays of his final usurpation, avalanches, torrents, rocks, and thunders, and above all these deadly glaciers, at once the proof and symbols of his reign;—add to this, the degradation of the human species—who in these regions are half deformed or idiotic, and most of whom are deprived of any thing that can excite interest or admiration. This is a part of the subject more mournful and less sublime; but such as neither the poet nor the philosopher should disdain to regard. (163)

Eric G. Wilson helpfully contextualises this passage in relation to Peacock's and Percy's interest in Zoroastrianism, which had been the inspiration for Peacock's unfinished poem *Ahrimanes.* At Bracknell in 1813, they had met the vegetarian thinker John Frank Newton, who had argued that 'humans, now overly civilized and thus separated from nature, were suffering under the reign of "Ahrimanes," the evil, dark, destructive deity in the dualistic Zoroastrian religion.'[24] Wilson argues that Percy's allusion to Ahriman, however, does not present him as 'a demonic magus to be overcome. On the contrary, as a personification of necessity, he is to be embraced.'[25] He goes on to claim that 'Mont Blanc' provides a way of embracing such cosmic forces by suggesting that the landscape is 'as much created in [the speaker's] own mind as it is discovered in his sight, for the impressions of his mind will one day produce the scene in the minds of others.'[26] While Wilson's overall argument here is plausible enough, he misreads the final sentence of the 'Ahriman' passage. The pronoun 'this' in the final sentence is not absolutely clear, but it seems unlikely to refer to 'Ahriman,' which is Wilson's claim. The Persian deity—whom we shall see also appears as an embodiment of powerful inhuman forces in *Manfred*—is for Percy clearly a figure of sublimity: a personification of Buffon's 'sublime but gloomy theory.' The 'more mournful and less sublime' subject, which should also be of concern to poets and philosophers is not Ahriman's supremacy, but 'the degradation of the human species.' Perhaps uncomfortable with Percy's racialised language and obvious disdain for the region's rural poor, critics have tended to ignore this part of the letter, but it is crucial to his understanding of environmental catastrophe and resonates with anxieties about the future of humanity in *Frankenstein*, 'Mont Blanc,' and 'Darkness.'[27] His concern with how global climate change—might cause the human race to degenerate suggests the influence of Buffon.[28] The naturalist had argued that quadrupeds of the Old World were bigger than those of America, where 'animated Nature is weaker, less active, and more circumscribed in the variety of her productions.'[29] He described Native Americans as physically, mentally, and sexually enervated, 'wandering savages' who had proved incapable of improving the landscape. The cause was a cold and humid climate in which 'every circumstance concurs in diminishing the action of heat.'[30] The implication of Percy's allusion is that global cooling—Ahriman's 'final usurpation'—would lead to the degeneration of all human beings to the level of North American or Swiss 'savages.' He therefore shows himself to drawing on the same discourse of climate improvement that was so influential on colonial administrators like Raffles. For Buffon, as Alan Bewell states, 'to be controlled by climate is to be closer to animals and plants than to civilized human beings.'[31]

Glacial augmentation, therefore, was not a topic of idle geological interest: it raised questions about the very future of humanity. Unsurprisingly, it is also a concern in other texts by the Diodati Circle that depict ice as a dynamic force. The 'Second Spirit' conjured by Manfred in Byron's poem describes how 'the Glacier's cold and restless mass / Moves onward day by day.'[32] Visiting the valley of Chamonix, Victor Frankenstein describes the Montanvert glacier as 'tremendous and ever-moving,' and, in the 1831 text, notes that it 'with slow pace is advancing down from the summit of the hills, to barricade the valley.'[33] The ill-fated protagonist of Polidori's novel, *Ernestus Berchtold; or the Modern Oedipus* (1819), states that whenever he is high in the Alps, he 'seems always to crouch some invisible being beneath whose power is infinite,' and which he feels he cannot 'resist':

> It seems that I hear him laughing audibly at our vain attempts to encroach upon his dominion. It appears to me as if the avalanche were but the weapon of his impatience, while he insidiously steals upon those habitations he has covered with his snows, by the silent, gradual approach of the glaciers.[34]

The influence of Shelley's letter to Peacock on this later passage is apparent, as is that of 'Mont Blanc,' which is printed a few pages after the mention of Buffon in the *History*.[35] The poem uses ice as a metonym for the sublime power of nonhuman forces, describing the 'glaciers' that 'creep':

> Like snakes that watch their prey, from their far fountains,
> Slow rolling on; there, many a precipice,
> Frost and the Sun in scorn of mortal power
> Have piled: dome, pyramid, and pinnacle,
> A city of death, distinct with many a tower
> And wall impregnable of beaming ice. (180)

Ice, here, has a kind of threatening agency. Percy's use of assonance and consonance—'gl**a**ci**er**s **cree**p', '**slo**w ro**ll**ing on'—emphasises the power of the glaciers through repetition while also slowing the movement of the verse to mimic the slow violence of global cooling. As William Keach points out, 'Mont Blanc' is remarkable in its 'crossing of extended blank-verse enjambment with irregular rhyme,' and the poem uses its rhymes to provide structure and shape.[36] The uncharacteristic aba rhyme of 'power' and 'tower' closely connects the terms, emphasising the scornful commentary that sublime and 'impregnable' glacial architecture—the rather Gothic 'city of death'—makes on more vulnerable human dwellings. The description of the ice as 'beaming' suggests not only its incandescent brightness, but also its expansive energies. The buildings 'piled' up by 'Frost and Sun' echo the 'palaces of death and frost' in the letter to Peacock, as well as 'the palaces of Nature, whose vast walls / Have pinnacled in clouds their snowy scalps' in Canto III of *Childe Harold's Pilgrimage*: the verbs 'piled' and 'pinnacled' both suggest natural agency (II, 100, ll. 591–592). There is a further echo in *Ernestus Berchtold*, in which the protagonist compares human and glacial architecture: 'the pyramids might rise unnoticed upon the rocks before my view [...] I cannot bear that human strength should be unable to stamp its hand on these towering memorials of convulsions we could not influence.'[37] The poem's sense of human weakness and insignificance in the face of the 'slow rolling on' of the glaciers coheres precisely with Percy's understanding of Buffon's views.[38] As Bewell puts it, 'Mont Blanc' 'describes a world in which the human power to create temperate environments seems impotent in the face of a power that dwells apart from human control.'[39]

The ice's destructive power is further emphasised as the poem shifts to a different metaphor, imagining it as 'a flood of ruin / [...] that from the boundaries of the sky / Rolls its perpetual stream'. This strange mingling of air and water suggests the glaciers' climatological force. In an image that will be repeated across a number of 1816 texts, the 'flood' leaves even 'vast pines' either 'strewing / Its destined path' or standing 'branchless and shattered.' Although the geomorphic capacity of glaciation was not fully understood in the early nineteenth century, Percy describes how

> [...] the rocks, drawn down
> From yon remotest waste, have overthrown
> The limits of the dead and living world
> Never to be reclaimed. (181)

The phrasing here is ambiguous, but seems to suggest the power of glaciers and the rocks that they transport (from the supposedly dead world that is beyond human time) to make the living world desolate—compare the 'desolating snows' in the letter to Peacock—and to break down 'the division between the two worlds.'[40] For their world is revealed to be not really 'dead' at all, but a world of nonhuman agents. The glaciers have the power to 'draw down' the rocks, which themselves 'overthrow' boundaries. These images of destruction lead the poem to contemplate the vulnerability of animals and humans in the face of this inevitable ('destined') process:

> [...] The dwelling-place
> Of insects, beasts, and birds, becomes its spoil;
> Their food and their retreat for ever gone,
> So much of life and joy is lost. The race
> Of man, flies far in dread; his work and dwelling
> Vanish, like smoke before the tempest's stream,
> And their place is not known. (181)

Since Jonathan Bate's influential book, *The Song of the Earth*, ecological approaches to Romantic writing have often emphasised its concern with place and (following Heidegger) the capacity of poetry especially to create 'a *revelation* of dwelling': a strong sense of humanity's connection to the nonhuman world.[41] And yet Romantic texts are often equally concerned with the difficulty or impossibility of dwelling within a strange or hostile environment. The destruction described here is physical and epistemological: the glacier's huge power (ironically imagined as a 'city') threatens humanity's sense of place and suggests its smoke-like evanescence as a species. 'Tempest's stream' echoes the mingling of air and water a few lines earlier and again suggests the climatological power of ice. The final line in the quotation is probably an allusion to Psalm 103:

> As for man, his dayes as grasse: as a flower of the field, so he flourisheth.
>
> For the winde passeth over it, and it is gone; and the place thereof shall know it no more.[42]

In both texts, humans lose their place in the world—they are blown by the wind like smoke or plants—due to longer-standing natural processes. And in both texts, the subject of 'known' is ambiguous. If people know landscapes, it seems that landscapes also know people, and if there is knowledge then there can always be forgetting. Percy's shift from the psalm's 'man' (meaning individual men) to the more communal 'race of man' suggests that he is going beyond simply observing the transience of individual lives. The unstoppable process of glaciation will erase the human species from the land.

Nonhuman animals also have an important role in this passage. By covering the pastures and forests, the glaciers efface the dwellings of 'insects, beasts, and birds,' leaving only an absence of 'life and joy.' Michael O'Neill suggests that the passage 'serves both to bind humans and animals into a shared ecosystem and to imply their separate destinies,' for the animals 'will presumably die' while humans may be able to flee somewhere else.[43] He argues that the comma after 'man' emphasises this separation; I would add that the full stop and caesura after 'joy is lost' is just as significant. He also argues that the repetition of 'dwelling' serves to emphasise the connection, to which I would also add the importance of the rhyme of 'place' (of animals) and 'race' (of humans) and then the repetition of 'place' (of humans). At this stage in the poem, the idea of a shared response to environmental change between the human and nonhuman is stronger than any sense of difference. The poem's engagement with anthropocentrism is, of course, philosophically complex, and I will return to the relationship between humans and nonhuman animals later in the chapter. For the moment, I want to focus particularly on the threat to human 'dwelling,' which connects 'Mont Blanc' strongly to *Frankenstein*, a text that is also deeply concerned with the difficulties of finding one's place in the world and the future of 'the race of man.'

Frankenstein, Global Cooling, and the Posthuman

Mary Shelley's novel is concerned with the impossibility of stable dwelling. Its restless movements between different parts of Europe and the Arctic mimic the Creature's restless and ultimately fruitless search for a location where he will be able to dwell unmolested, as well as Victor's restless desire to shape his environment. As in her husband's work, the locale around Mont Blanc provoked Mary to consider the fragility of human communities. After the trauma of William's murder by the Creature and Justine's unjust execution, the Frankenstein family hope to find recuperation in 'an excursion to the valley of Chamounix' (121). One morning, Victor awakes to depressingly bad weather, reminiscent of the 'dreary night of November' when he brought the Creature to life in what is described as a 'catastrophe' (84). He seeks solace in a lone trek to the summit of Montanvert, on the northern slopes of Mont Blanc. 'The presence of another would destroy the solitary grandeur' of the landscape, he notes,

> It is a scene terrifically desolate. In a thousand spots the traces of the winter avelanche may be perceived, where trees lie broken and strewed on the ground [...] The path, as you ascend higher, is intersected by ravines of snow, down which stones continually roll from above; one of them is particularly dangerous, as the slightest sound, such as even speaking in a loud voice, produces a concussion of air sufficient to draw destruction upon the head of the speaker. (123)

Like other Alpine tourists of the period, Victor seeks an encounter with the sublime. But this version of the experience does not involve the safe contemplative distance that is often associated with the ability to aestheticize a potentially dangerous landscape. As befits Victor's risk-taking character, he emphasises the physical danger affecting the human observer. Given the emphasis on the sublime power of *silence* in 'Mont Blanc,' it is significant that the 'slightest sound' is dangerous in this setting: the silence necessary to avoid causing an avalanche also emphasises the alienness of the landscape. And, as in Percy's poem, the threat to human dwelling is represented metonymically through the destruction of the trees. Along with glacial augmentation, this is a key trope in the repertoire of the Diodati Circle in 1816. It is most starkly presented by Byron, who was always looking for symbols of self-ruination. The entry for 23 September 1816 in his 'Alpine Journal' describes how he 'passed *whole woods of withered pines—all withered*—trunks stripped and barkless—branches lifeless—done by a single winter—their appearance reminded me of me & my family.'[44] The journal was kept for Byron's beloved half-sister Augusta Leigh, a prototype of Manfred's lost Astarte. Early in that poem, Manfred contemplates suicide on the Jungfrau and compares himself, 'grey-hair'd with anguish' to 'these blasted pines, / Wrecks of a single winter, barkless, branchless, / A blighted trunk upon a cursed root' (IV, 64; I.ii.66–68). Both passages emphasise what has been lost—ecologically and emotionally—by turning nouns into negative adjectives through the addition of the suffix '-less'; a key technique, as we shall see, in 'Darkness,' a text which also contemplates the destruction of forests as linked to the fate of humanity.

The image of the 'strewing' and 'stripping' of the pine trees by the glacier's power connects 'Mont Blanc,' *Frankenstein*, *Manfred*, the 'Alpine Journal,' and the Shelleys' journals from the period. In the third journal-letter to Peacock in the *History of a Six Weeks' Tour*, Percy describes the Boisson glacier in the Chamonix valley as 'the most vivid image of desolation that it is possible to conceive. [...] The pines of the forest, which bound it at one extremity, are over-thrown and shattered to a wide extent at its base' (159–160). Similarly, in Mary's journal for 24 July 1816—when she also makes the first mention of composing *Frankenstein*—she writes that

> Nothing can be more desolate than the ascent of this mountain — the trees in many places have been torn away by avelanches and some half leaning over others intermingled with stones present the appearance of a vast & dreadful desolation.[45]

The Shelleys' prose accounts of the pines emphasise the scene's superlative 'desolation'; that is, its barrenness and its lack of inhabitants. 'Desolate' derives from the Latin *desolatus*, meaning 'left alone': this is a landscape in which human communities cannot flourish. This inhospitality is further emphasised in Mary's account as the travellers try to ascend Montanvert, but are forced to retreat:

> It began to rain almost as soon as we left our inn — when [we] had mounted considerably we turned to look on the scene — a dense white mist covered the vale & tops of scatered pines peeping above were the only objects that presented themselves — The rain continued in torrents — we were wetted to the skin so that [...] we resolved to turn back.[46]

One might see the white mist as marking the boundary between the living and the dead worlds described in 'Mont Blanc.' The travellers succeed in making the ascent the following day and find themselves, with familiar hyperbole, in 'the most desolate place in the world.'[47] This place is undoubtedly alien; 'Mont Blanc,' as we shall see, considers the extent to which such an environment can be comprehended by human consciousness. However, *Frankenstein* presents the possibility of a new species arising that will be more comfortable in Buffon's frozen environments.

After ascending Montanvert, Victor walks across the glacier to the opposite mountain, so that he has a view of Mont Blanc 'in awful majesty' (124). The individual apotheosis potentially offered by the sublime—the swelling of the heart provoked by a 'wonderful and stupendous scene'—is cut short by his sudden encounter with the Creature, their first since his creation (124). Victor's sublime solitude is interrupted by a painful reminder of the communal responsibilities that we have to each other, but also of his inferiority to his creation, who travels with 'superhuman speed' and 'easily eludes' Victor's attempt to attack him physically. The Creature responds to his creator's loathing with counter threats, but also with justifications, pointing out his natural benevolence, his loneliness, and Victor's failure to care for him. Therefore, he states,

> The desert mountains and dreary glaciers are my refuge. I have wandered here many days; the caves of ice, which I only do not fear, are a dwelling to me, and the only one which man does not grudge. These bleak skies I hail, for they are kinder to me than your fellow-beings. (126–127)

The inhospitable landscape where humans and animals cannot 'dwell,' and which for Victor is a kind of touristic site, is for the Creature the safest dwelling place that he can find. Cast out by his creator, and by the human communities to which his sensibility, if not his appearance, should connect him, the Creature has been forced to find a different sort of connection by living within the apparently 'dead' world of rocks and glaciers and finding a form of fellowship (as the punning word 'hail' suggests) even with the 'bleak' weather.

The anthropologist Julie Cruikshank has contrasted a Western colonial idea of glaciers as 'pristine, wild, and remote from human influence' with the views of indigenous populations in Alaska and the Yukon, who (she suggests) see them as 'intensely *social* spaces where human behaviour, especially casual hubris or arrogance, can trigger dramatic and unpleasant consequences in the physical world.'[48] Whereas for Victor the glacier is a sublime backdrop to his anthropocentric imagination, for the Creature it offers the possibility of a relationship with the world. Despite his apparently 'unnatural' beginnings, the Creature is therefore shown here and at other points in the text to be more connected to nonhuman nature than the human characters. He can flourish in any environment, but has to tell his story to Victor in a constructed dwelling: a mountain hut with a fire and therefore at a temperature suitable for his creator's 'fine sensations' (127). The state of nature can be presented in idyllic, Rousseauvian terms, as when he promises that he 'will go to the vast wilds of South America' with his mate and live a vegetarian existence: 'we shall make our bed of dried leaves; the sun will shine on us as on man and will ripen our food. The picture I present to you is peaceful and human' (170). But writers of the Romantic period were increasingly aware of another state of nature: a state of rapid, violent, and uncontrollable change.[49] The encounter between Victor and the Creature in the vale of Chamonix connects the Creature's agency to that of the glaciers. This agency is often destructive, although the allusion to 'Kubla Khan' (the 'caves of ice' within the Khan's 'pleasure-dome') in the Creature's speech may suggest a creative power as well. After all, in 'Mont Blanc,' the glaciers not only destroy dwellings, but also construct 'dome, pyramid, and pinnacle'. Furthermore, beneath this 'flood of ruin':

> [...] vast caves
> Shine in the rushing torrent's restless gleam,
> Which from those secret chasms in tumult welling
> Meet in the vale, and one majestic River,
> The breath and blood of distant lands, for ever
> Rolls its loud waters to the ocean waves,
> Breathes its swift vapours to the circling air. (181–182)

The allusions to 'Kubla Khan' are palpable, but whereas in Coleridge's poem, the river that bursts forth from 'that deep romantic chasm' eventually sinks 'in tumult to a lifeless ocean,' here the glaciers are seen as part of a global hydrological cycle upon which life is dependent. It is the Creature's potential to create new life that Victor finds most disturbing.

Just as Percy extrapolated from the augmentation of a single glacier to consider the icy fate of the whole world, the Creature's power and resilience generates reflections on the future of humanity as a species. His appearance offers an ironic commentary on Victor's fantasy on the way to Chamonix of 'the mighty Alps, whose white and shining pyramids and domes towered above all, as belonging to another earth, the habitations of another race of beings' (121). In a recent discussion of *Frankenstein* and human extinction, Claire Colebrook follows earlier political readings by seeing the Creature as 'a disenfranchised other who could, in theory, be redeemed and included.'[50] She places Mary with thinkers like Marx, Adorno, and Jameson, who seek political solutions to what may seem an intolerable existence: 'what appears to be existentially unacceptable should be transformed through social and political revolution. If recognition were granted to the potential hordes of the future one would be faced not with violence but with sympathy and pity.'[51] Colebrook seems to see such thinking as narrow and utopian, given humanity's impact on the environment: 'the question is not one of how we humans can justify hostile life, but how we can possibly justify ourselves given our malevolent relation to life.'[52] As Colebrook recognises, political critique is crucial to the novel; nonetheless, I think she underestimates its willingness to face up to the prospect of human extinction, rather than seeing it as a problem to be solved. The Creature may present to Victor a vision of a 'peaceful and human' existence in South America with his mate (170), but this is a rhetorical ploy; to view the Creature as no more than a surrogate for disenfranchised humans is to miss precisely what is important about the novel in the context of Romantic geotheory and the Diodati Circle's concern with human-environmental interactions. Victor eventually destroys his work on the Creature's companion due to his fear that the two might procreate: 'a race of devils would be propagated upon the earth, who might make the very existence of the species of man a condition precarious and full of terror' (190). Victor assumes that this new posthuman species would be far more resilient than humans and would therefore destroy them; having read Buffon, he is well aware of the adaptive value of being able to survive in colder climates.[53] It is therefore entirely fitting that the novel begins and ends in the Arctic. After being responsible for the deaths of Victor's family, the Creature leads him 'to the everlasting ices of the north' so that he will suffer further privations: 'you will feel the misery of cold and frost, to which I am impassive' (227). *Frankenstein* raises the spectre of a posthuman future in which a new species develops that is able to flourish on Buffon's icy globe.[54]

As Siobhan Carroll has recently shown, Arctic exploration in the Romantic period was enmeshed with debates around climate change and geoengineering projects. She argues convincingly that 'situating *Frankenstein* in a climatological context enables us to see the ramifications of Victor's experiment as symptomatic of a larger cultural concern over Europeans' readiness to wield the nature-shaping power of imperial science'.[55] The relationship between knowledge and power is a key theme in the novel and evident in how Walton and Victor try to impose their wills on to the world. One of Walton's goals is to find the fabled Northwest passage linking the Atlantic to the Pacific. Such a discovery would have been of enormous benefit to British imperialism and was, of course, the aim of a number of government-supported expeditions in the nineteenth century. Walton's 'voyage of discovery' through 'pathless seas' (58, 317) is an attempt at reterritorialization: by beginning with an entirely deterritorialized idea of the Arctic as a blank space without indigenous human inhabitants, or nonhuman creatures, or elemental agencies, he presents the possibility that it can be mastered cartographically as well as physically, so that he may achieve 'the sight of a part of the world never before visited, and may tread a land never before imprinted by the foot of man' (50). His fantasy is of a land 'where cold and frost are banished' (49), but the Arctic itself proves resistant to imaginative projections: '*Frankenstein* sounds the death knell of dreams of the undiscovered *terra nullius* in its depiction of Walton's defeat in polar space.'[56] This resistance is most powerfully apparent in the cold reality that 'immures' his ship among threatening 'mountains of ice' (234) and forces Walton and his crew to turn back when the ice eventually breaks up. Imperialism seeks to impose itself onto the environment through writing the landscape as possession. However, in Mary's novel ice is a force recalcitrant to any such process. Furthermore, their experience of the Alpine landscape and Buffon's speculations suggested to the Shelleys that the empire of ice itself had the capacity to overwrite all traces of the human.

The novel's concern with Arctic exploration is inflected by its relationship to 'The Rime of the Ancyent Marinere.'[57] In his second letter to his sister, Walton notes that 'I am going to unexplored regions, to "the land of mist and snow"; but I shall kill no albatross, therefore do not be alarmed for my safety' (55).[58] It is likely that Mary was working from the *Lyrical Ballads* version of the poem, although it had been reprinted, with the addition of Coleridge's marginal gloss, in *Sibylline Leaves* (1817). Its depiction of human vulnerability in the face of extreme weather conditions must have resonated particularly during a period of climate crisis: 'And now the storm-blast came, and he / Was tyrannous and strong.'[59] Like *Frankenstein*, Coleridge's poem is particularly concerned with the power of ice over humans:

> Listen, Stranger! Mist and Snow,
>   And it grew wond'rous cold:
> And Ice mast-high came floating by,
>   As green as emerald.
>
> And thro' the drifts the snowy clifts
>   Did send a dismal sheen;
> Ne shapes of men ne beasts we ken—
>   The Ice was all between.
>
> The Ice was here, the Ice was there,
>   The Ice was all around:
> It crack'd and growl'd, and roar'd and howl'd—
>   Like noises of a swound![60]

This is another 'desolate' space, where there is no place for humans or animals: another space defined by absence ('ne shapes [...] we ken'). In such an environment, the ice itself is given an uncanny agency through anaphora and personification. Like Mont Blanc's glaciers, it is not static, but threatens through its unpredictable movements. The final simile seems to suggest that it resembles a person experiencing some sort of violent fit ('swound'). Coleridge's poem, like *Frankenstein*, presents the failure and vulnerability of humanity's imperialising aspirations when faced with uncooperative objects. Both texts have been subject to dubious moralising interpretations by ecological writers. Bruno Latour finds in Mary's novel the injunction 'love your monsters': 'Dr. Frankenstein's crime was not that he invented a creature through some combination of hubris and high technology, but rather that he ***abandoned the creature to itself***'.[61] He sees the novel as offering a 'parable for political ecology' by showing that human beings need to take responsibility for the entanglements of their technologies with the natural world. *Frankenstein* is therefore co-opted in line with ecomodernist discourse, which calls for a 'good Anthropocene' created by a doubling down on the human capacity to shape the planet.[62] A more traditional ecological moral is evident in James McCusick account of Coleridge's poem: 'by blessing the water-snakes, the Mariner is released from his state of alienation from nature [...] [He] has learned what the Albatross came to teach him: that he must cross the boundaries that divide the natural world, through unmotivated acts of compassion between "man and bird and beast".'[63] One might well sympathise with these readings, but they each have the two same problems. First, they focus on the relationship between humans and nonhuman creatures and ignore the significance of recalcitrant objects to the moral framework of both texts. Secondly, they do not reflect the complexity of their sources, and particularly the way in which both texts are concerned with fear, abjection, disgust, and the deep difficulties of connecting with others. Rather than offering straightforward moral messages, *Frankenstein* and the 'Rime' are focused on the potential incompatibility of human aspirations and desires with the dynamic environmental processes upon which we are dependent. In different ways, Latour and McKusick offer a inadequately anthropocentric responses to texts that, whatever else they do, tend to emphasise the precarity and finitude of the human.

Humans, Animals, Elements

The relationship between ecological precariousness and problems of knowledge and navigation is also important in Byron's 'Darkness.' The poem has a range of influences, from the climate of 1816, to various apocalyptic passages in the Bible, to Enlightenment thinkers including Buffon, to the European sun-spot panic of the same year.[64] Although it is unlikely that Byron knew anything about the Tambora eruption, his vision of the Earth descending into utter blackness also resonates strongly with accounts of the effects of the ash cloud in Raffles's narrative. But the poem's imagining of the heat-death of the universe, and particularly of human extinction, without any form of eschatological recompense moves far beyond its intertexts to produce a vision of nihilistic horror. Towards the end of *Nihil Unbound: Enlightenment and Extinction*, Raymond Brassier offers a reading of an essay by Jean-François Lyotard that considers the philosophical implications of the future death of the sun.[65] As Brassier states:

> The extinction of the sun is a catastrophe, a mis-turning or over-turning (*kata-strophe*), because it blots out the terrestrial horizon of future possibility relative to which human existence, and hence philosophical questioning, have hitherto oriented themselves.[66]

There is a sense, therefore, in which imagining the death of the sun means that human extinction *has already happened*: not in the sense of 'the termination of a biological species,' but in that it destroys any idea of human transcendence by revealing our contingency and superfluousness: 'if the extinction of the sun is catastrophic, this is because it disarticulates the correlation' (that is, the idea that reality can only be understood in terms of the human relationship to it).[67] Furthermore, because the 'time of extinction' entails '*the extinction of space-time* [...] it is not so much that extinction *will* terminate the correlation, but that it *has already* retroactively terminated it.'[68] The paradoxical temporality of extinction is reflected in Byron's poem. The speaker has a prophetic 'dream, which was not all a dream' (IV, 40, l.1) and as a result is able to describe the destruction of all life on earth and the universal triumph of Darkness in the past tense, as if it has already happened. The poem simultaneously affirms human exceptionality—our unique power to imagine our own absence—while presenting our destruction alongside that of all other lifeforms and forces in the universe.

> The bright sun was extinguish'd, and the stars
> Did wander darkling in the eternal space,
> Rayless, and pathless, and the icy earth
> Swung blind and blackening in the moonless air. (IV, 40, ll. 2–5)

This describes on a cosmic scale the future desolation that the Shelleys imagined when they contemplated the advancing glaciers in the valley of Chamonix. In Byron's poem, the planet becomes unable to sustain human life; as the words 'rayless', 'pathless', and 'moonless' suggest, the poem is defined by absence, loss, and confusion. The image of the earth swinging blindly through the air suggests a movement that deviates from its normal orbital trajectory. The usual order of the universe has collapsed; the darkening of the sun is mirrored by the other stars, which now wander without a clear path. Ultimately leaving the planet 'lifeless,' the dwindling of the sun leads initially to anthropogenic environmental destruction:

> And the thrones,
> The palaces of crowned kings—the huts,
> The habitations of all things which dwell,
> Were burnt for beacons; cities were consumed
> [...]
> Forests were set on fire—but hour by hour
> They fell and faded—and the crackling trunks
> Extinguish'd with a crash—and all was black. (IV, 41, ll. 10–13, 19–21)

As in 'Mont Blanc,' the dwellings of humans and nonhumans—'all things' suggests both—are destroyed, although in this case the environmental factors are mediated through human action. And, once again, the destruction of forests, in this case by fire rather than ice, stands for the fate of humanity. In this black world, volcanos—those old symbols of human vulnerability—become places of shelter: 'Happy were those who dwelt within the eye / Of the volcanos, and their mountain-torch' (IV, 41, ll. 16–17). Buffon's 'Epochs of Nature' had described the importance of volcanos to early human development and cultivation of the land:

> To destroy the brushwood and the forests, they would employ the flames derived from volcano's [sic], or from their burning lavas; for, with the assistance of this powerful element, they cleared and purified the grounds which they chose to inhabit.[69]

'Darkness' shows humanity regressing to a primitive state in which the technology of fire no longer has any purpose, for there is nothing left to burn. As Buffon had imagined it had begun, human civilisation ends by depending on volcanic energy.

Like 'Mont Blanc,' 'Darkness' addresses the effects of a cooling world on humans and nonhumans. People wander in the 'despairing light' in horror, anger, and madness. Normal animal activity is disrupted:

> The wild birds shrieked
> And, terrified, did flutter on the ground,
> And flap their useless wings; the wildest brutes
> Came tame and tremulous; and vipers crawl'd
> And twined themselves among the multitude,
> Hissing, but stingless—they were slain for food. (IV, 41, ll. 32–7)

Significantly, the effects of the catastrophic environmental change on animals are as much behavioural as physical. The birds' wings are 'useless' because of fear. Losing their ability to navigate the air, they mimic the pathless stars and the blindly swinging Earth in the poem's opening lines and become prone and vulnerable. The behaviour of the vipers can be read as the kind of strange portent that might accompany a Biblical apocalypse, but it can also be read ecologically. In the absence of the sun's rays, cold-blooded creatures are unable to maintain their body heat. Therefore they seek to 'twine' themselves among humans for warmth. In a poem full of cruel ironies, this leads to their destruction. 'Darkness' shows birds losing their defining characteristic of flight and serpents losing their defining characteristic of venom. (That there are such things as flightless birds and venomless snakes does not detract from the symbolic power of the idea.) 'Stingless' is yet another negative adjective created from a noun, in a text littered with them. This loss of species-being may also explain why the vipers are drawn to humans; their intertwining with the 'multitude' suggests how the darkness erases the differences between creatures.

With the planet deprived of the sun's energy, lack of food causes bloody conflicts around resources. 'War… / Did glut himself again,' and

> All earth was but one thought—and that was death,
> Immediate and inglorious; and the pang
> Of famine fed upon all entrails—men
> Died, and their bones were tombless as their flesh;
> The meagre by the meagre were devoured,
> Even dogs assail'd their masters, all save one,
> And he was faithful to a corse. (IV, 41–42, ll. 38, 42–48)

Citing this passage, Wood argues that 'with a remarkable, prescient sympathy, Byron's "Darkness" anticipates the full-blown humanitarian disaster as it was to unfold in Switzerland and around the world over the subsequent three-year global climate emergency.'[70] The bad weather of 1816 certainly led to a terrible subsistence crisis in Europe. Switzerland, which already had many inhabitants who were barely surviving, was particularly affected; in 1817, the price of grain almost tripled. It is indeed remarkable that Byron and the Shelleys write so forcefully about environmental catastrophe in the early stages of the Tambora crisis, without of course knowing about the volcanic eruption the previous year. However, I am not sure that 'sympathy' is the right word to use for any of their accounts. The Shelleys' approach to the poor Swiss whom they encounter in the *History* is generally disdainful rather than sympathetic. And Byron[71] generally treats human suffering in 'Darkness' with a kind of horrified irony. In the above passage, famine paradoxically feeds, men are reduced to 'tombless' flesh and bones, and the starving consume the starving, presumably with little success. Environmental catastrophe leads not to stronger communities—whether among humans or between humans and nonhumans—but a brutally Hobbesian universe. Even dogs—supposedly humanity's most loyal companion—turn out to be another species that lose their defining characteristic: in this case, by turning on their masters. The one 'faithful' dog, who protects his master's body from animal and human predators until he himself dies of starvation, does not offer an ethical model but is rather an exception who proves the general rule of selfish predation. The mode of the poem is sceptical rather than sympathetic. Perhaps its most terrifying aspect is the way in which it condenses a period of global cooling that Buffon had imagined taking place over millennia, so that civilised rules and ethics, and whole ecosystems, are shown to collapse in the speed it takes to read a few lines of text. Beginning with the blindly wandering planet and stars of the poem's opening lines, much of the poem gives a feeling of chaotic and futile energy that dissipates as we arrive at the final few lines. They continue the emphasis on loss and absence from the beginning of the text—'Seasonless, herbless, treeless, manless, lifeless' (IV, 43, l. 71)—but also present a new sense of stillness after the death throes of all human, creaturely, and even elemental agencies:

> The waves were dead; the tides were in their grave,
> The moon their mistress had expired before;

> The winds were withered in the stagnant air
> And the clouds perish'd. (IV, 43, ll. 78–81)

Given our increasing understanding of the speed with which climate change can take place—and the extent to which natural variation can be escalated by human action—the poem's powerful compressions seem much more prescient than any sympathy.[72]

In a suggestive discussion of *The Last Man*, Kate Rigby argues that the novel subverts 'patriarchal and anthroparchal assumptions': 'the demise of Man (as defined in accordance with the logic of colonization), it is hinted, might just open the way for the emergence of a new kind of human-nonhuman collectivity.'[73] The question of human exceptionalism is also of considerable importance to the texts of 1816. We have seen that 'Mont Blanc' suggests at one point that humans and nonhumans are similarly affected by the destruction of dwelling, although, as I will discuss below, it also asks whether human beings give the universe meaning in a way that other creatures do not. *Frankenstein* suggests how an obsession with improving the human race could lead to its demise. It is notable that Victor prefers bleak, sublime environments denuded of flora and fauna; his only significant encounter with nonhumans are the 'living animal[s]' (82) that he vivisects in order to discover the secrets of life. In contrast, the Creature enjoys pastoral landscapes and is often fascinated by 'the pleasant song of the birds' (130), which he tries to imitate. The utopian life in the New World that he imagines with his mate suggests a potential *post*human-nonhuman collectivity: 'my food is not that of man; I do not destroy the lamb and the kid, to glut my appetite' (170). There is a significant parallel here with Byron's *The Prisoner of Chillon* (1816), a poem that draws on the familiar tropes of extinction and ruination to contemplate the fate of a prisoner who sees his brothers die and is the last of a 'failing race.'[74] Bonnivard's solitary anguish is simultaneously subjective and cosmic in scope: 'There were no stars—no earth—no time / […] / But silence, and a stirless breath' (IV, 10, ll. 245–247). But the end of the poem describes a new form of community that has emerged from loss and devastation:

> With spiders I had friendship made,
> And watch'd them in their sullen trade,
> Had seen the mice by moonlight play,
> And why should I feel less than they?
> We were all inmates of one place,

And I, the monarch of each race,
Had power to kill—yet, strange to tell!
In quiet we had learn'd to dwell—
My very chains and I grew friends
So much a long communion tends
To make us what we are:—even I
Regain'd my freedom with a sigh. (IV, 16, ll. 381–394)

Despite the horrors of his confinement, Bonnivard is eventually able to turn his cell into a dwelling place through the relationships that he builds, not only with nonhuman animals, but with the objects that confine him ('my very chains'). In contrast to some of Byron's other verse of 1816, identity here emerges not from the solitary striving of the sublime individual but from a 'communion' with apparently insignificant entities. One might read the poem's ending as ironically bathetic in that it reveals the depth of Bonnivard's fall through the lowliness of his connections and the fact that he has become institutionalised to the extent that he regrets regaining his freedom. But, in part due to the first-person perspective, the tone here is very different from that of 'Darkness': empathetic rather than ironic.

'Darkness,' in contrast, is sceptical about any form of collectivity and some other Byron poems of 1816 also suggest the desire to separate human and nonhuman worlds. The tragedy of the human, as explained in *Manfred*, is that we are caught between the creaturely and the spiritual: 'half dust, half deity, alike unfit / To sink or soar' (IV, 63, l. 40). And in 'Prometheus,' 'Man is in part divine / A troubled stream from a pure source' (IV, 32, ll. 47–48). In Canto III of *Childe Harold's Pilgrimage*, absorption in the sublime landscape offers a temporary respite from this painfully antithetical state:

I live not in myself, but I become
Portion of that around me; and to me
High mountains are a feeling, but the hum
Of human cities torture: I can see
Nothing to loathe in nature, save to be
A link reluctant in a fleshly chain,
Class'd among creatures, when the soul can flee,
And with the sky, the peak, the heaving plain
Of ocean, or the stars, mingle, and not in vain. (II, 103, ll. 680–688)

This stanza splits human experience into the elemental-spiritual and the creaturely-physical. The speaker's claim that he loathes 'nothing' in nature is undercut by his professed desire to break the 'fleshly chain.' There is a kind of litotes at work here: by seeking to escape from the physical body, and therefore from all connection with other humans and animals, he actually seems to loathe a great deal. 'Nature,' here, describes elemental forces and objects rather than biological entities. A key question for the Diodati Circle was the extent to which human beings should be 'class'd among creatures.' Does a human being have much in common with a dog or a bird or a serpent, or does the power of the imagination—and associated technologies—make humanity a cosmic, elemental force that transcends 'creatureliness'?

The Romantic sublime was not just manifest in a desire to dissolve the self into the nonhuman world, but also in a desire to shape and control that world. In 'Epochs of Nature,' Buffon had suggested that global cooling might be at least temporarily delayed, or even reversed, by human cultivation of the earth: 'the draining, clearing, and peopling a country will give it a warmth which will continue for some thousand years.'[75] He adduces several examples to show that:

> Man can have an influence on the climate he inhabits, and, in a manner, fix its temperature at any point that may be agreeable to him; and, what is singular, it is more difficult for him to cool than to heat the earth. He is master of the element of fire; which he can augment and propagate at pleasure, but not of the element of cold, which he can neither lay hold of or communicate.[76]

Buffon's surprising confidence in the salvific technology of fire is treated ironically by Byron and Mary Shelley.[77] In 'Darkness,' the destruction of the forests stands for the destruction of humanity, as it does in 'Mont Blanc,' but the agents of the destruction in Byron's poem are humans themselves who are engaged not in enlightened cultivation, but the last desperate act of an imminently doomed species. In *Frankenstein*, Waldman celebrates 'the new and almost unlimited powers' of natural philosophers: 'they can command the thunders of heaven, mimic the earthquake' (76). Victor's enthusiasm first catches fire at the age of fifteen when he witnesses 'a most violent and terrible thunder-storm' emerging from 'behind the mountains of Jura.' This scene was no doubt inspired by the weather of 1816, for it echoes Mary's account in the *History* of the thunder storms at the start of June that I quoted from at the beginning of this chapter and in which she describes 'observing the lightning play among the clouds in various parts of the heavens, and dart in jagged figures upon the piny heights of Jura.' One particularly sublime storm 'lit up' Lake Geneva, 'when a pitchy blackness succeeded, and the thunder came in frightful bursts over our heads amid the darkness' (99–100). Byron, of couse, would have seen the same storms and in Canto III of *Childe Harold's Pilgrimage* imagines the 'live thunder' leaping among 'the rattling crags,' producing a sort of call-and-response between Jura and the Alps (II, 110, ll. 164–165). In both Byron's poem and *Frankenstein*, the solitary subject is thrilled by the sublime weather and desires its power, to the extent that Byron's narrator fantasises about throwing

> Soul, heart, mind, passions, feelings, strong or weak,
> All that I have sought, and all that I seek,
> Bear, know, feel, and yet breathe—into *one* word,
> And that one word were Lightning. (II, 112, ll. 908–911)

This moment of supreme creativity offers a conflation of subjective and elemental agency, therefore briefly overcoming the idea of human precarity that haunts the Diodati Circle's texts.

In *Frankenstein*, a bolt of lightning 'utterly' destroys 'an old and beautiful oak': a 'catastrophe' that thrills Victor (69–70). This key term connects the destruction of the oak tree to the destruction of Victor's family, the 'catastrophe' of the Creature's creation (84), and the Creature's self-immolation. Victor's Promethean spark is shown as ultimately destructive rather than creative. The Creature, too, uses fire for destructive purposes, most obviously when he burns the De Laceys' cottage after he eventually realises that it is not a place in which he will ever be able to dwell (in a passage that echoes the burning of the 'huts' in Byron's poem):

> As the night advanced, a fierce wind arose from the woods, and quickly dispersed the clouds that had loitered in the heavens: the blast tore along like a mighty avelanche, and produced a kind of insanity in my spirits, that burst all bounds of reason and reflection. I lighted the branch of a

> tree, and danced with fury around the devoted cottage. [...] With a loud scream, I fired the straw, and heath, and bushes, which I had collected. The wind fanned the fire, and the cottage was quickly enveloped by the flames, which clung to it, and licked it with their forked and destroying tongues. (163)

As Andrew Griffin notes, 'the whole cosmos seems to conspire in [the Creature's] revenge.'[78] The passage develops the Creature's earlier assertion of his 'fellow-feeling' with the weather by suggesting not only that it reflects his mood but also that it aids his plans. The wind is a cruel mixture of fire and ice: a 'mighty avelanche' that fans the flames. With grim irony, this passage presents the Creature as a profoundly elemental force whose gifts of firewood earlier in his relationship with the De Laceys have been transformed into a destructive sacrifice. At this point in the novel, the Creature embodies those aspects of nature that prove to be beyond human dominance and control and suggests the limits of the Promethean vision of humanity as inevitably improved by the communication of fire or other technology. Indeed, in the same chapter the Creature even compares himself to an avalanche: 'finding myself unsympathized with, [I] wished to tear up the trees, [and] spread havoc and destruction around me' (161). The novel is therefore akin to 'Mont Blanc', which—in its vision of the glacier's 'scorn' for human power—recalls Buffon's insistence on the recalcitrance of 'the element of cold' to human control or transmission.

The relationship between the human and the elemental is also profoundly important to *Manfred*, concerned as it is with the 'Mysterious Agency' of nonhuman nature (IV, 54; I.i.28). In the first scene of the drama, the protagonist seeks to 'compel' the 'spirits of the unbounded Universe' to his will: 'the Promethean spark, / The lightning of my being' (IV, 58; I.i.154–155). He eventually succeeds through 'a tyrant-spell, / Which had its birthplace in a star condemn'd, / The burning wreck of a demolish'd world' (IV, 54; I.i.43–45). Somehow, out of cosmic destruction emerges Manfred's earth-shattering power, but that power is also shown to be profoundly dangerous to its wielder. He summons seven spirits, of 'earth, ocean, air, night, mountains, winds, [and] thy star' (IV, 57; I.i.132). The latter seems to be the earlier 'star condemn'd', for its ruling spirit describes how before the earth was even created, it existed as 'a world as fresh and fair / As e'er resolved round sun in air; / Its course was free and regular' (IV, 57; I.i.112–114. Now, however, it has become

A wandering mass of shapeless flame,
A pathless comet, and a curse,
The menace of the universe;
Still rolling on with innate force
Without a sphere, without a course.
A bright deformity on high,
The monster of the upper sky! (IV, 57; I.i.117–123)

The star's amorphousness is part of its threat; like other hyperobjects, it challenges the idea that the human imagination can shape the world. In its vision of a deformed star wandering 'pathless' through the universe, *Manfred* draws richly on the etymology of the word 'disaster,' as does the image of the 'pathless' stars and 'blind and blackening' earth in 'Darkness.' The term connects Manfred's earthly power and misfortune to the cosmic realm: a dis-aster is a star that has gone bad and portends suffering on earth. These longstanding links between the human and the sidereal are given new force in the Romantic period by a more scientific concern with the role of the human in relation to huge scales of time and space. To what extent, the play asks, does the imagination have any significance in relation to vast, cosmic forces? One possible answer can be found in the dramatic form itself, which can be read as a form of mental theatre in which all the characters, human and nonhuman, are merely projections of Manfred's psychodrama. But, as Timothy Morton points out, the play hesitates between this solipsistic view and a dramatic framing through which 'theatrical space itself, an analogue for the environment of the Alpine setting, surrounds and negates Manfred's power.'[79] Thus it is appropriate that the spirit of Manfred's monstrous birth-star simultaneously obeys and scorns the magician and suggests that his apparent mastery is temporary: 'forced by a power (which is not thine, /And lent thee but to make thee mine)' (IV, 57; I.i.126–127).

Later in the poem, Manfred refuses to bow down to Arimanes, the master of the spirits, and yet he needs his power to summon the ghost of Astarte. Arimanes, like Ahriman in Percy's letter to Peacock, embodies all the volatile forces of the cosmos and their lack of care for humanity:

Hail to our Master!—Prince of Earth and Air!—
  Who walks the clouds and waters—in his hand
The sceptre of the elements, which tear
  Themselves to chaos at his high command!
He breatheth—and a tempest shakes the sea;
  He speaketh—and the clouds reply in thunder;
He gazeth—from his glance the sunbeams flee;
  He moveth—earthquakes rend the world asunder.
Beneath his footsteps the volcanoes rise;
  His shadow is the Pestilence; his path
The comets herald through the crackling skies;
  And planets turn to ashes at his wrath.
To him War offers daily sacrifice;
  To him Death pays his tribute; Life is his,
With all its infinite of agonies—
  And his the spirit of whatever is! (IV, 81; II.iv.1–16)

In contrast to the unrhymed pentameter of Manfred's Shakespearean soliloquies, this is the rhymed language of worship and incantation. The use of anaphora, caesurae, archaic verb forms, and exclamations creates a mode of address that is both formal and full of unruly energy. 'Life' is defined by the Spirits who celebrate Arimanes as a set of catastrophes and traumas—'an infinite of agonies'— and he causes cosmic destruction through the most basic aspects of his being: breathing, speaking, gazing, and moving. He may be a grouping of elements, but the inclusion of 'War' also suggests the entanglement of human and nonhuman forces. Arimanes is a catastrophic assemblage and his capacity to destroy planets links him to the 'demolish'd world' that is the source of Manfred's power.

Like Victor Frankenstein and Childe Harold, Manfred prefers the 'Wilderness' denuded of other beings: 'where the birds dare not build, nor insect's wing / Flit o'er the herbless granite' (IV, 72; II.ii.64–5). If he is unfortunate enough to encounter another person, he feels himself 'degraded back to them' (IV, 73; II.ii.78). There is a tension in the play between Manfred's identification with other humans—'We are the fools of time and terror' (IV, 76; II.ii.164)—and his resistance to the idea of being classed among any living creatures. There is a similar tension in Byron's 'Alpine Journal,' which relishes its human encounters but also seeks to replace them with landscapes: 'I have lately repeopled my mind with Nature.'[80] In the 'Journal,' this aspiration is unachievable, with Byron lamenting at the end that he has been unable to 'lose my own wretched identity in the majesty & the power and the Glory.' *Manfred*[81] is more celebratory of the power of sublime selfhood. For Andrew Hubbell, Manfred's 'attempt to commune with nature is actually self-communing, part of his quest to separate himself from all other "breathing" creatures, a quest based on a dualist understanding of nature as dead matter animated by (his own) dominating spirit.'[82] One might therefore read the play in ecologically moralising terms, as suggesting the ways in which the human capacity and desire to harness the elements threatens a dangerous denial of the reality that we are part of the natural world and vulnerable to elemental forces. But the arc of the text resists any such straightforward conclusion. Famously, in the final deathbed scene, Manfred refuses his apparently Faustian fate by sheer force of will, refuting the claim that his 'power' was temporarily gifted to him by the spirits: rather, it was 'purchased by no compact with thy crew / But by superior science' (IV, 101; III.iv.113–15). Like Byron's 'Prometheus,' Manfred makes 'Death a Victory' through his defiance of cosmic forces and his assertion of the power of the human mind to stand against the brute contingency of the universe (IV, 33, l. 59). It is a pyrrhic victory, of course, for while he will not kneel to Arimanes, he prostrates himself to his 'own desolation'; nonetheless, this seems to me the most anthropocentric of all the texts produced by the Diodati Circle in 1816. It presents human destinies, however ill-starred, as profoundly different to those of nonhuman creatures and heroises Manfred's refusal to bow down to the elements.

'Mont Blanc' and the Problem of Extinction

This chapter concludes by returning to the topic of deep time in 'Mont Blanc,' a poem which offers the richest example of the Diodati Circle's meditation on anthropocentrism and extinction. To use modern philosophical language, it is a text about correlationism, which Quentin Meillassoux defines as 'the idea according to which we only ever have access to the correlation between thinking and being, and never to either term considered apart from the other.'[83] The poem worries away at the question of the extent to which the universe means anything beyond human consciousness. In an influential account, Earl Wasserman argues that Shelley (to summarise crudely) generally adopts a sceptical version of Berkeleyan idealism that grounds reality not in the mind of God but in a 'universal Mind.'[84] If Shelley adheres to 'a metaphysics that eternalizes the Self or the Mind, turning the latter into the perennial mirror for the manifestation of the entity' then, following Meillassoux, he cannot be a correlationist in the strict sense.[85] However, critics have also connected 'Mont Blanc' suggestively to the Kantian sublime, with its emphasis on the power of consciousness, and to Kant's attempt to bridge the apparent dichotomy between materialism and idealism.[86] Other scholars have emphasised instead Shelley's debt not to the idealism of Berkeley or Kant but to 'a sceptical epistemology derived from David Hume and William Drummond.'[87] Cian Duffy even forcefully argues, partly against Wasserman, that the poem's 'ontology is unequivocally empirical and materialist.'[88] It is beyond the scope of this chapter to give an account of Percy Shelley's complex and shifting philosophical views, or fully to address the ontology or epistemology of 'Mont Blanc': in any case, it seems to me helpful to read it as a poem of tentative questioning rather than philosophical dogma. This chapter attempts to understand the poem's epistemological questions as also ecological ones, as indeed they were for Kant. Nigel Clark has suggestively analysed Kant's separation of human beings from nonhuman nature in the context of the sense of vulnerability engendered by environmental catastrophes such as the Lisbon earthquake, about which the philosopher wrote a seismological treatise: 'beneath the exaggerated claims of autonomy, the often repressed sensuous and somatic aspects of selfhood, the desire to lord it over nature, lies Kant's intuition of bodily vulnerability, an awareness that humankind is both individually and collectively bared to its elements.'[89] This provides a useful way of thinking about the oscillating shifts between mind and universe in 'Mont Blanc.' For Percy Shelley, addressing epistemology also means addressing the fate of the human species in a potentially hostile universe. According to Clark, Kant 'sought to adequate the regularities and potentialities of the universe with those powers that defined our species being, only to find, again and again, that the magnitudes of the geo-cosmic side of the equation defied containment in the categories provided by the self-willing subject.'[90] I want to suggest that 'Mont Blanc' can be accurately described in similar terms, and self-consciously reflects on its own failures to 'contain' the nonhuman. In the Preface to the *History of a Six Weeks' Tour*, Percy describes the poem as an attempt to 'imitate' the 'untameable' and 'inaccessible' (6). Its knowingly quixotic attempt to imagine a desolate landscape separate from human perception is also an attempt to imagine a universe existing on a time scale that reduces human beings to transient phenomena.

Much of the sublimity in 'Mont Blanc' comes not from the vast size or obscurity of the mountain, but from the speaker's sense of the locale's connection to deep time. The pines around the Arve are described as 'children of elder time' and the sound that they create while 'swinging' in the wind as 'an old and solemn harmony.' We have seen that, for the Diodati Circle, pine trees could stand metonymically for humans in their mutual vulnerability to more powerful forces, but these sturdier pines stand for the prehuman past. The 'Ravine of Arve,' the addressee of this section, manifests 'its own deep eternity,' a phrase on the verge of tautology that is difficult to understand in the 1817 text in the *History*:

> Thine earthly rainbows stretched across the sweep
> Of the ethereal waterfall, whose veil
> Robes some unsculptured image; the strange sleep
> Which when the voices of the desart fail
> Wraps all in its own deep eternity. (177)

The first main clause describes the play of light and water in the ravine; 'unsculptured' may have both physical and epistemological meanings, referring to a rock face that has been carved out by natural rather than human processes, and also a landscape that does not depend on human perception for its existence.[91] The second part of the quotation is more difficult, in part because the semicolon suggests a relationship between the 'strange sleep' and the ravine's waterfalls and rainbows, but the relationship is not explained.[92] The Scrope Davies Manuscript has a version of the passage that helps in some respects:

> [...] even the sleep,
> The sudden pause that does inhabit thee
> Which when the voices of the desart fail
> And its hues wane, doth blend them all and steep
> Their periods in its own eternity.[93]

The 'strange sleep,' it is now clear, is a 'pause' when the landscape's 'unresting sound[s]' are silent, or at least inaudible, and 'its hues wane.' The editors of the Longman edition suggest that this describes 'the interval of suspense when the sound is interrupted and the view [of the

Arve] obscured by an obstacle.'[94] This seems a little simplistic, however, because it hardly helps to make sense of the final clause. The sleep is a 'pause' that somehow manifests an 'eternity' that includes the 'periods' of sound and colour. At this point, the poem seems to be in danger of collapsing under its own paradoxes. The landscape's sounds do not rest and yet they can be paused. The sleep is temporary but also eternal. 'Periods,' as pointed out in the Longman edition, refers both to 'their times of existence' and 'their times of cessation.'[95] For all its brilliance, 'Mont Blanc' is a poem in which ambivalence and complexity sometimes tips over into incoherence, a fact that has not always been accepted by the poem's critics.

However, it helps to connect that 'strange sleep' to 'the trance sublime and strange' a few lines later. Rather than caused simply by a material blockage in sight and sound, the sleep is a pause in consciousness when the external world fades and flattens, and the perceiving subject turns inward. Whether one thinks of the poem as concerned here with individual or universal mind, the implication is that consciousness simultaneously transcends and incorporates ('blends') the temporary 'periods' (however long) of 'the universe of things,' even the 'elder time' of pine trees and especially rocks. The human apprehension of deep time is celebrated as an epistemological folding in of the universe within human perception. 'Eternity,' it is implied, is ultimately a product of the human imagination. One of the crucial effects of this sublime apprehension is that the present becomes 'ethereal' and insubstantial. The speaker moves into 'a trance sublime and strange,' which apparently breaks down subject and object into an 'unremitting interchange':

> One legion of wild thoughts, whose wandering wings
> Now float above thy darkness, and now rest,
> Where that or thou art no unbidden guest,
> In the still cave of the witch Poesy,
> Seeking among the shadows that pass by
> Ghosts of all things that are, some shade of thee,
> Some phantom, some faint image; till the breast
> From which they fled recalls them, thou art there! (177)

Wasserman suggests that 'along the walls of the cave, as in Plato's myth, pass sensory images, [...] not the things-in-themselves, which the human mind cannot know.' He argues that the trance ultimately confirms that 'reality is neither the subjective impression nor the external thing, but the active and irresolvable mental tension between the two that is embodied in the word "Seeking."'[96] This insight supports an ecological approach to the poem, which would see Mont Blanc as a kind of 'hyperobject,' a thing that is 'massively distributed in time and space relative to humans' and that can be apprehended only partially by human perception and is always retreating from it. As Morton describes it, 'hyperobjects are not simply mental (or otherwise ideal) constructs, but are real entities whose primordial reality is withdrawn from humans.'[97] Mont Blanc is much more than the topographical feature perceived by the poem's speaker: 'any "local manifestation" of a hyperobject is not directly the hyperobject.'[98] The phrase 'ghosts of all things that are' suggests that the effect of the human apprehension of eternity—the mind's ability to contain the 'periods' of the landscape—is to make the everyday phantasmagorical. But this is not necessarily a fully idealist position, as Morton explains: 'ecological existence is with ghosts, strangers, and specters, precisely because of reality, not in spite of it. [...] An object fails to coincide with its appearance-for another object, no matter how accurate that appearance-for.'[99] Mont Blanc is a hyperobject particularly in its temporal 'nonlocality'; the 'ghosts' of the present-day landscape evoke a deep and haunting geological history.[100] One might read the whole poem as an attempt to understand the reality or otherwise of the history—a human narration—of a period that is outside human consciousness. It addresses what Meillassoux identifies as a key problem for modern correlationist philosophy: 'the conditions of meaning for dia-chronic statements [...] about events that are anterior or ulterior to every terrestrial-relation-to-the-world.'[101]

The following section of the poem moves from the potential solipsism of the 'still cave' to an outward-looking gaze. As Wasserman states, 'the ability of thought to float above that darkness [of the ravine] into the transcendent is now symbolized by the dramatic gesture of the poet's raising his glance above the ravine to the snow-covered peak of Europe's highest mountain.'[102] In addition, the transcendent properties of the mind are symbolised by its capacity to imagine a world in which humans are absent:

> A desart peopled by the storms alone,
> Save when the eagle brings some hunter's bone,
> And the wolf tracts her there—how hideously

Its shapes are heaped around! rude, bare, and high,
Ghastly, and scarred, and riven.—Is this the scene
Where the old Earthquake-demon taught her young
Ruin? Were these their toys? or did a sea
Of fire, envelope this silent snow?
None can reply—all seems eternal now. (179)

Frances Ferguson notes how, despite its sublime subject, the poem 'projects an air of sociability.'[103] The paradoxical phrase 'voices of the desart' is repeated in the 'desart peopled' by storms. The non-anthropocentric sociability is emphasised further by the connection between nonhumans (the eagle and the wolf) and 'peopling.' In an ironic twist, the only human imprint on this 'peopled' landscape is in the form of an archaeological relic, intimating how deep time succeeds as well as precedes the human. The suggestion that this apparently deserted place has its own 'voices' should be read alongside the description in the *History* of the Shelleys' journey towards Geneva and the Alps—'the natural silence of that uninhabited desert contrasted strangely with the voices of the men who conducted us' (93)—as well as the passage in *Frankenstein* in which the human voice is seen as dangerous in such a 'desolate' environment because it may cause avalanches (123). 'Mont Blanc' imagines a form of community that includes the perceiving subject and the various forms of the landscape—and potentially supplants connections with human communities—much as Byron declares in the 'Alpine Journal' that 'I have lately repeopled my mind with Nature.'[104] The landscape appears 'eternal,' but perhaps it is only the 'eternity' of the human mind in a moment of sublime transcendence that can comprehend its catastrophic changeability.[105] The mind achieves this through a form of sociable projection, imagining what Ferguson calls a 'domestic circle' featuring the 'Earthquake-demon' and her young.[106] The rhetorical questions suggest both the uncertainties of early nineteenth-century geology—was the landscape created by an earthquake or a volcanic eruption?—and the stimulatingly confusing effect that deep time has on the human subject (and which is most palpable in the complex rhetorical question that ends the poem). They also suggest a desire for connection with this ever-receding landscape: for the mountain to answer back.

At the end of the section, the 'wilderness' is imagined to speak with a 'mysterious tongue.' The poem asserts that the mountain has 'a voice [...] to repeal / Large codes of fraud and woe.' Ferguson's important essay on 'What The Mountain Said' has surprisingly little to say about this one moment when the mountain is given the power of speech (although we do not hear from it directly), suggesting only that its capacity to repeal the 'large codes' comes from 'making it clear that a love of humanity is easy if one can love a mountain that is physically inimical to men.'[107] More recently, Louise Economides has read this passage as key to the poem's critique of anthropocentrism and its concern with 'the contours of nature as a material force that exceeds human communication.' She finds in the mountain's voice 'an absence that can only negate, repealing discursive fraud but not grounding new forms of linguistic stability.'[108] Neither reading, it seems to me, really addresses the political claims being made at this point in the poem. As far as I am aware, Cian Duffy and Nigel Leask are the only critics to have addressed its politics of catastrophe. Leask suggests that Shelley is working against the mobilisation of catastrophe by 'counter-revolutionary apologists to figure the characteristic manifestation of divine providence, intervening in natural and human history as an agent of castigation.'[109] Following Hutton and others, Shelley 'set out to show catastrophe as part of a *self*-regulating (as opposed to *divinely* regulated) economy.'[110] Leask's apparent assumption that less homeostatic geotheories than Hutton's were providential can certainly be challenged; Cuvier's geological 'revolutions,' for example, had little to do with any divine plan. However, he is right to draw attention to the relationship between the mountain's evocation of deep time and its moral and political meanings. The history of environmental change may emphasise human vulnerability, but it also challenges conservative Christian readings of the landscape such as Coleridge's 'Hymn before Sun-Rise, in the Vale of Chamouni' (1802), to which Shelley was in part responding.[111] As Duffy argues, the poem offers an alternative and politically radical view available to the 'cultivated imagination' by understanding 'Nature' as 'subject not to divine regulation, but to its own internal (Necessary) laws.'[112] Similarly, in the 'Hymn to Intellectual Beauty,' the 'frail spells' of religion and superstition are rejected in favour of a more philosophical response to the sublime that may 'free / This world from its dark slavery'.[113] The politics of both poems can be connected to Ahriman and 'the adamantine hand of Necessity,' as described in the letter to Peacock: a terrifying force whose 'avalanches, torrents, rocks, and thunders, and above all these deadly glaciers' threaten to destroy humanity. But Ahriman's 'usurpation' is also a revolution against providential understandings of the universe. In

Shelleyan terms, Mont Blanc and the catastrophic necessitarian power that it represents should be understood as closer to the frightening but liberating Demogorgon of *Prometheus Unbound* (1820) rather than the oppressive and despotic Jupiter. That this can only be understood by a small elite of 'the wise, and great, and good' is entirely in keeping with the tendency of the Diodati Circle's texts of 1816 to place intellectuals far above other groups of people, such as the supposedly brutish Swiss peasantry or the apparently empty-headed tourists, including the English woman whom Byron mocks in the 'Alpine Journal' for experiencing Chamonix as 'rural' rather than sublime.[114] As in the case of the narrative of the eruption discussed in the previous chapter, Byron and the Shelleys' response to the Tambora crisis must be understood as a perspective only possible for an elite protected from its more severe effects by rank, wealth, and education.

Mont Blanc offers a moral and political lesson precisely because it shows the contingency of all human understandings of the universe and our vulnerability to the power of Necessity. As a hyperobject, it is simultaneously available and unavailable to the speaker's consciousness, experienced as 'remote, serene, and inaccessible.' These qualities separate it not only from the speaker, but from other aspects of the environment:

> The fields, the lakes, the forests, and the streams,
> Ocean, and all the living things that dwell
> Within the daedal earth; lightning, and rain,
> Earthquake, and fiery flood, and hurricane
> [...]
> The works and ways of man
> [...]
> All things that move and breathe with toil and sound
> Are born and die; revolve, subside, and swell. (179–180)

This is a difficult passage to relate to the rest of the poem, in part because the main verb ('are') is delayed until the twelfth line. Before then, we get an assemblage that includes topographical features, weather, geological phenomena, seasons, and humans, all of which are shown to be subject to the transience that 'Power,' embodied by the mountain, rises above.[115] Here 'eternity' lies separately from the human mind *and* from the ancient landscape. The apparent contrast between human life and the events of deep time is briefly collapsed, for environmental phenomena, even those responsible for the earth's geomorphology, seem to be included as part of the list of 'things that move and breathe.' 'Daedal earth' is a conventional poetic phrase suggesting the world's richness and variety; 'daedal' can also suggest cunning artifice.[116] But if the earth is wrought, it can also be unwrought by a 'remote' and disinterested Power, and the gap between humanity and deep time is reasserted later in the fourth section where human dwelling is described as threatened by perpetually expanding glaciers. Earlier in this chapter, I discussed the poem's emphasis on the destructive power of ice, but noted that the very end of the fourth section also presents it as a creative force in that it feeds the 'majestic' Arve—'The breath and blood of distant lands' (181)—as part of the hydrological cycle: 'Rolls its loud waters to the ocean waves / Breathes its swift vapours to the circling air' (182). For Duffy, this shows the poem moving beyond 'the conventional catastrophic account of the natural sublime' to comprehend the 'long-term benefits' of apparent destruction. He reads this move in political terms, with the creativity of Mont Blanc's glaciers 'showing the ultimate triumph of revolutionary values.'[117] This is plausible enough in relation to his excellent general discussion of the politics of the sublime in Shelley. But his reading of the glaciers in relation to 'Volneyan ruins of empire tropes' ignores the fact that 'Mont Blanc' is not only concerned with human history.[118] Certainly natural and political histories are intertwined in the poem, as they are throughout Shelley's work, but when 'Mont Blanc' is read in relation to the other 1816 writings discussed in this chapter, it also raises the possibility that natural history supersedes that of the 'race of man.' In Meillassoux's terms, the poem understands elemental forces not only as *ancient* (distant in time), but as *ancestral* (anterior or ulterior to any human relationship to the world).[119]

The complexities of 'Mont Blanc's' fourth section show how difficult it is to make sense of the poem's understanding of the relationship between humans, mutability, and eternity. Frances Ferguson argues that the poem 'creates an image of sublimity that continually hypostatizes an eternity of human consciousness.' She understands the poem in line with post-Kantean correlationism: 'because even the ideas of the destructiveness of nature and the annihilation of mankind require human consciousness to give them their force, they thus are testimony to the necessity of the continuation of the human.'[120] This seems a little too neat for such a troubled text. Certainly it is not clear that the poem goes so far as to assert the 'necessity' of human continuation, and the reference to the 'secret strength of things / Which governs thought' (182) in the final section suggests a greater interest in the power of materiality over human consciousness than Ferguson's reading allows for. This section presents a landscape denuded of human perception, in which 'none beholds' the descending snows, 'Winds contend / Silently', and the lightning is 'voiceless' (182). The poem then, famously, turns things on their heads by asking what sort of meaning they have outside human experience:

> And what were thou, and earth, and stars, and sea,
> If to the human mind's imaginings
> Silence and solitude were vacancy? (183)

As we have seen, Ferguson finds in the poem 'an implicit argument for the transcendent existence of man—not because man is able to survive the threat posed by the power of the material world, but because he is able to domesticate the material world as Mont Blanc from an object into a found object.'[121] Charles E. Robinson reads the ending in similar terms: 'if the human mind did not interpret—did not make meaningful and moral—the powers of nature, then these energies would remain insignificant—vacant, barren.'[122] Other critics have understood it more ambivalently. Wasserman sees it as an expression of Shelley's scepticism about the possibility of certitude in empirical or imaginative experience.[123] I.J. Kapstein suggests that the lines, despite their ambiguity, form an 'anti-climax' that contradicts the assertion of the mountain's power over the mind earlier in the poem.[124] More recently, Duffy has argued that 'the poem's final lines re-affirm the central sceptical point of Shelley's revision of the discourse of the sublime: only the "wise" imagination can go beyond the defeat of the understanding towards an intuition of the Necessity informing the landscape.'[125] And Economides reads the question as 'compellingly open rather than rhetorical because it acknowledges that some aspects of the domain we call "nature" may well exceed our socio-linguistic constructions'; that is, it asks the reader whether silence should be equated with vacancy.[126] Despite the richness and sophistication of these critical accounts, they seem to me to miss something important by focusing solely on individual epistemology. Clearly the poem asks profound epistemological questions, but what if we were to read it in relation to the future of 'the race of men' rather than the perception of the individual mind? After all, at the end Shelley is not only imagining a particular place desolate of humankind, but *the whole universe*. One might read the ending as arrogantly anthropocentric in its implication that the world without us is nothing but 'vacancy.' But the poem as a whole manifests a more complex engagement with the paradoxes inherent in human representations of a world without humans. It therefore, like 'Darkness,' relates to one of the key tensions in Anthropocenic thinking: the ways in which we imagine ourselves as haunting a future from which we are simultaneously absent (extinct) and present (as writers of the stratigraphic record and through the very act of imagining). As Colebrook observes, 'how might we imagine a world without organic perception, without the centered points of view of sensing and world-oriented beings? [...] Can we imagine a mode of reading the world, and its anthropogenic scars, that frees itself from folding the earth's surface around human survival?'[127] The poem does not offer a solution to these questions, or to Meillassoux's related problem of diachronicity (that is, how we are able to make statements about the world as it exists separate from our relationship to it). Rather, and with remarkable prescience, it tests the limits of language's capacity to move between 'the dead and living world' and explores different perspectives on the significance of the human species to the universe: as vulnerable and transient victims of environmental flux, as co-creators participating in an 'unremitting interchange,' and as transcendental subjects without whom there would be only 'vacancy.'[128]

In all its contradictions and capaciousness, 'Mont Blanc' addresses the key ecological and philosophical ideas that run through the texts that I have examined in this chapter. Like 'Darkness' and *Frankenstein*, it explores the precarity of human dwelling on an ever-changing planet, and imagines the possibility of a posthuman universe. At the same time, it also imagines the transcendental power of the human mind, as it is celebrated (ambivalently) in *Manfred* and *Childe Harold's Pilgrimage*, and (partly) ironised in *Frankenstein*. For all its speculative force, the poem is also deeply grounded in the empirical experience of the Swiss landscape and the 'Year without a Summer' that we find captured in the *History of a Six Weeks' Tour* and the letters and journals of the Diodati Circle. Taken as a whole, these texts suggest a deep connection between assertions of the human capacity to transcend the material world and an apprehension of human vulnerability due to the 'radical asymmetry of the relationship between human existence and nature.'[129] Ice is a particularly powerful symbol of this asymmetry; it stands metonymically for the epistemological 'withdrawnness' of objects and the existence of a '*glacial* world [...] capable of subsisting without any of those aspects that constitute its concreteness for us.'[130] There is certainly a utopian strand in Romantic writing about environmental change, but for these authors in 1816, standing in the twin shadows of Tambora and Mont Blanc, the human species seemed unlikely to resist the violence of deep time. However bleak this view may seem, it also offers a valuable counter to what Bronislaw Szerszynski describes as the 'soteriological dream of security' offered by 'a calculative technological response' to weather and climate.[131] Even those texts that emphasise human power over the elements hardly offer a stable sense of salvation, for such power is shown to be temporary, dangerous, and isolating. The intertextual connections that are so crucial to all of these works offer a potentially positive communal response through sympathetic interchange to the problems of human vulnerability. But this is an elite community cut off from the majority of the human race, who are presented as unable to transcend their mere creatureliness: most powerfully, in the brutal state of nature depicted in 'Darkness.' The perils of such an exceptionalist perspective are suggested by the threat that Frankenstein's Creature represents to humanity as a whole. The precarity and finitude of the human species, detached from any eschatological narrative, is investigated with remarkable prescience in these texts, as is the capacity of apparently enlightened humans to exacerbate the destructive power of natural forces.

Notes

1. [Mary Shelley and Percy Bysshe Shelley], *History of a Six Weeks Tour* (London: T. Hookham and C. and J. Ollier, 1817), 96–97. Further references to the *History* are in the text. Until she married at the end of 1816, Mary was known as Mary Wollstonecraft Godwin, but I have referred to her throughout this chapter as (Mary) Shelley, as this is how she is known as an author. In discussing the Shelleys, I have done my best to move between forenames and surnames so that it is always clear to which one of them I am referring. Percy Shelley is referred to as Shelley throughout the final section of the chapter, as it focuses on 'Mont Blanc'.
2. Shelley and Shelley, *History*, 99.
3. Gillen D'Arcy Wood, *Tambora: The Eruption that Changed the World* (Princeton: Princeton University Press, 2014), especially Chap. 3. A key earlier intervention is by Jonathan Bate, who discusses 'Darkness' and

'To Autumn' in relation to Tambora in *The Song of the Earth* (London: Picador, 2001), Chap. 4. Bate was not the first critic to discuss the importance of Tambora to the Diodati Circle: see John Clubbe's unduly neglected article, 'The Tempest-toss'd Summer of 1816: Mary Shelley's *Frankenstein*," *Byron Journal* 19 (1991): 26–40. A groundbreaking account of 'Mont Blanc' in relation to Tambora can be found in Alan Bewell, *Romanticism and Colonial Disease* (Baltimore: The Johns Hopkins University Press, 1999), Chap. 6.

4. Earlier studies on textual relations between these authors include William D. Brewer, *The Shelley-Byron Conversation* (Gainesville: University Press of Florida, 1994); Charles E. Robinson, *Byron and Shelley: The Snake and Eagle Wreathed in Flight* (Baltimore: The Johns Hopkins University Press, 1976); and Robinson's 'Byron and Mary Shelley and *Frankenstein*' (University of Nottingham: Byron Centre, 2000), accessed 26 March 2016, http://byron.nottingham.ac.uk/resources/digital/foundation%20lectures/Robinson.pdf.
5. Dipesh Chakrabarty, 'The Climate of History: Four Theses,' *Critical Inquiry* 35 (2009): 197–222.
6. Lowell Duckert, 'Glacier,' *postmedieval: a journal of medieval cultural studies* 4 (2013): 68–79 (71). Duckert is quoting Jane Bennett, *Vibrant Matter: A Political Ecology of Things* (Durham and London: Duke University Press, 2010), 54. Fascinating meditations on ice, agency, culture, and climate change can be found in Julie Cruikshank, *Do Glaciers Listen? Local Knowledge, Colonial Encounters, and Social Imagination* (Vancouver: UBC Press, 2005) and Susan Oliver, 'Ice Thoughts 2', 9 November 2016, accessed 20 January 2017, https://susanoliverweb.com/2016/11/09/ice-thoughts-2/.
7. As noted in the introduction, these terms are contested and I am not suggesting that either describes a monolithic philosophical 'movement.'
8. Quentin Meillassoux, *After Finitude: An Essay on the Necessity of Contingency*, trans. Raymond Brassier (London: Bloomsbury, 2008), 5; Raymond Brassier, *Nihil Unbound: Enlightenment and Extinction* (Basingstoke: Palgrave, 2007), 229–230; Timothy Morton, *Hyperobjects: Philosophy and Ecology After the End of the World* (Minneapolis: University of Minnesota Press, 2013), 1.
9. For a fine analysis of the relationship between Romantic literature and geology, see Noah Heringman, *Romantic Rocks, Aesthetic Geology* (Ithaca and London: Cornell University Press, 2004).
10. Martin Rudwick, *Bursting the Limits of Time: The Reconstruction of Geohistory in the Age of Revolution* (Chicago: University of Chicago Press, 2005), 140.

11. Rudwick, *Bursting*, 142–149. According to Thomas Jefferson Hogg, Percy had read Buffon in 1811 and even 'carefully' translated a 'treatise' by the naturalist: see *The Life of Percy Bysshe Shelley*, 4 vols (London: Edward Moxon, 1858), I, 458. Given that most earlier parts of the *Histoire Naturelle* had been translated into English twenty years earlier by William Smellie, this may well have been the 'Les époques de la nature.' While recognising the importance of this text, Smellie had not provided a full translation of it on the basis that it was 'perhaps too fanciful to receive the general approbation of the cool and deliberate Britain': see Buffon, Georg-Louis Leclerc, *Natural History, General and Particular*, trans. William Smellie, 9 vols (London: A. Strahan and T. Cadell), 3rd edition, IX, 258. Mary Shelley's reading list for 1817 includes 'Buffon's theorie du terre,' which outlined his first 'steady-state' geotheory (and not, as some critics seem to think, his theory of global cooling): see *The Journals of Mary Shelley, 1814–1844*, ed. Paula R. Feldman and Diana Scott-Kilvert, 2 vols (Oxford: Clarendon Press, 1987), I, 100. It is not clear whether Mary had read Buffon's theories about global cooling separately or only knew of them via Percy.
12. See Martin Rudwick, ed., *Georges Cuvier, Fossil Bones, and Geological Catastrophes: New Translations & Interpretations of Primary Texts* (Chicago and London: The University of Chicago Press, 1997). Cuvier's catastrophism is an important influence on *Cain* (1821). For useful accounts of Cuvier and the later Romantics, see Melissa Bailes, 'The Psychologization of Geological Catastrophe in Mary Shelley's *The Last Man*,' *ELH* 82 (2015): 671–699 and Christine Kenyon Jones, '"When this world shall be *former*": Catastrophism as imaginative theory for the younger Romantics,' *Romanticism on the Net* 24 (2001), 16 paragraphs, accessed 22 March 2016, http://www.erudit.org/revue/ron/2001/v/n24/006000ar.html.
13. Nigel Leask, 'Mont Blanc's Mysterious Voices: Shelley and Huttonian Earth Science,' in *The Third Culture: Literature and Science*, ed. Elinor S. Shaffer (Berlin and New York: Walter de Gruyter, 1998), 182–203.
14. Rudwick, *Bursting*, 158–172.
15. For the philosophical impact of the Lisbon earthquake, see Nigel Clark, *Inhuman Nature: Sociable Life on a Dynamic Planet* (London: Sage, 2011), Chap. 4, and Susan Neiman, *Evil in Modern Thought: An Alternative History of Philosophy* (Princeton: Princeton University Press, 2002).
16. Clark, *Inhuman Nature*, 93.
17. Rudwick, *Bursting*, 233.
18. Eric G. Wilson, *The Spiritual History of Ice* (Basingstoke: Palgrave, 2003), 93.

19. Buffon had suggested in 'Epochs of Nature' that the Alpine glaciers '*constantly augment in all their dimensions*' (IX, 363; italics in original). Certainly, glacial expansion around Chamonix had been a serious threat to the local inhabitants in the seventeenth and early eighteenth centuries: see Brian Fagan, *The Little Ice Age: How Climate Made History 1300–1850* (New York: Basic, 2000), 123–127.
20. The original letter does not include the nineteen words after 'mass of frost'; *The Letters of Percy Bysshe Shelley*, ed. Frederick L. Jones, 2 vols (Oxford: Clarendon Press, 1964), I, 499. It is likely that Mary Shelley added them to the *History of a Six Weeks' Tour*, possibly to emphasise the importance of polar ice to Buffon's theory. See Siobhan Carroll, 'Crusades Against Frost: *Frankenstein*, Polar Ice, and Climate Change in 1818,' *European Romantic Review* 24 (2013): 211–30 (227).
21. Leask, 'Mysterious Voices,' 195.
22. Heringman also takes issue with Leask on this point (*Romantic Rocks*, 76).
23. See Eric Gidal, '"O Happy Earth! Reality of Heaven!": Melancholy and Utopia in Romantic Climatology,' *Journal for Early Modern Cultural Studies* 8 (2008): 74–101.
24. Wilson, *Spiritual History*, 97. See also Timothy Morton, *Shelley and the Revolution in Taste: The Body and the Natural World* (Cambridge: Cambridge University Press, 1994), 65–67.
25. Wilson, *Spiritual History*, 105–106.
26. Wilson, *Spiritual History*, 106.
27. An exception to this critical neglect is Noah Heringman, who reads the passage in the context of Shelley's critique of the idealisation of Swiss mountaineers: see *Romantic Rocks*, 77–79.
28. The passage transcribed by Mary Shelley in the *History* omits the final sentence of this paragraph in the original letter: 'It presents views, a development of which I reserve for conversation' (*Letters*, I, 499). This tantilising statement suggests that Percy was genuinely interested in racial degradation and wanted to consider it further. For evidence of Mary's transcription, see her *Journal*: 'In the evening I copy S.'s letter to Peacock' (I, 117). Percy's concern with climate and degeneration is apparent in *A Vindication of Natural Diet* (1813), which connects humanity's adoption of meat-eating and consequent problems with 'some great change in the climates of the earth': *The Prose Works of Percy Bysshe Shelley*, ed. E.B. Murray, 1 vol. (Oxford: Clarendon Press, 1993), I, 77.
29. Buffon, *Natural History*, V, 115.
30. Buffon, *Natural History*, V, 133. Buffon's ideas were popularised by writers such as William Robertson: see Anya Zilberstein, *A Temperate*

Climate: Making Climate Change in Early America (Oxford: Oxford University Press, 2016), 37–38.

31. Alan Bewell, 'Jefferson's Thermometer: Colonial Biogeographical Constructions of the Climate of America,' in *Romantic Science: The Literary Forms of Natural History*, ed. Noah Heringman (New York: SUNY Press, 2003), 111–138 (122).
32. Lord Byron, *The Complete Poetical Works*, ed. Jerome McGann, 7 vols (Oxford: Clarendon Press, 1980–1993), IV, 55 (I.i.68–69). Further references to Byron's works are to this edition and are in the text.
33. Mary Shelley, *Frankenstein*, ed. D.L. Macdonald and Kathleen Scherf, 2nd edn (Peterborough, Ontario: Broadview, 2005), 123, 338. Further references to *Frankenstein* are in the text.
34. John William Polidori, *Ernestus Berchtold; or the Modern Oedipus. A Tale* (London: Longman, 1819), 12.
35. Unless otherwise noted, I am using the version of 'Mont Blanc' originally published in the *History.*
36. William Keach, *Shelley's Style* (Methuen: New York and London, 1984), 196.
37. Polidori, *Ernestus Berchtold*, 12.
38. In two other versions of these lines, 'mortal power' is 'human power': *The Complete Poetry of Percy Bysshe Shelley*, ed. Donald Reiman and Neil Fraistat, 3 vols (Baltimore: The Johns Hopkins University Press, 2000–2012), III, 518.
39. Bewell, *Romanticism and Colonial Disease*, 224.
40. Shelley, *Complete Poetry*, III, 519.
41. Bate, *Song*, 266.
42. Psal. CIII. 15–16. The quotation perhaps also alludes to Nahum III. 17 and Job VII. 10. Shelley, *Complete Poetry*, III, 520.
43. Shelley, *Complete Poetry*, III, 520.
44. *Byron's Letters and Journals*, ed. Leslie A. Marchand, 12 vols (London: John Murray, 1973–1982), V, 102.
45. Byron, *Journals*, I, 117–118.
46. Byron, *Journals*, I, 118.
47. Byron, *Journals*, I, 119.
48. Cruikshank, 10–11.
49. Building on Susan Neiman's work, and with particular attention to Kant, Clark (*Inhuman Nature*, Chap. 4) relates the development of the autonomous 'modern subject' during the period to anxieties about human vulnerability to environmental change.
50. Claire Colebrook, *Death of the PostHuman: Essays on Extinction, Vol. 1* (Michigan: Open Humanities Press, 2014), 195.

51. Colebrook, *Death*, 196.
52. Colebrook, *Death*, 198.
53. Before attending university, Victor attends an incomprehensible lecture and therefore 'became disgusted with the science of natural philosophy, although I still read Pliny and Buffon with delight, authors, in my estimation, of nearly equal interest and utility' (70). This passage does not appear in the 1831 edition. Carroll points out that Victor fears the Creature because it is 'well-suited to inheriting a frozen world' ('Crusades', 222).
54. Mary Shelley would of course return a decade later to the theme of the total destruction of humanity in *The Last Man*, a text that is also deeply concerned with climate change: see Bewell, *Romanticism and Colonial Disease*, Chap. 9; Carroll, 'Crusades', 222; Kate Rigby, *Dancing with Disaster: Environmental Histories, Narratives, and Ethics for Perilous Times* (Charlottesville and London: University of Virginia Press), Chap. 2.
55. Carroll, 'Crusades,' 220. See also her book *An Empire of Air and Water: Uncolonizable Space in the British Imagination, 1750–1850* (Philadephia: University of Pennsylvania Press, 2015), 51–55. Other useful accounts of *Frankenstein* and Arctic exploration include Adriana Craciun, 'Writing the Disaster: Franklin and *Frankenstein*,' *Nineteenth-Century Literature* 65 (2011): 433–480 and Jessica Richard, '"A Paradise of My Own Creation": *Frankenstein* and the Improbable Romance of Polar Exploration,' *Nineteenth-Century Contexts* 25 (2003): 295–314. See also Wood, *Tambora*, Chap. 6, for a discussion of the warming of the Arctic in the context of Tambora.
56. Carroll, *Empire*, 54.
57. For a useful reading of this relationship, see Beth Lau, ed., *Fellow Romantics: Male and Female British Writers, 1790–1835* (Farnham: Ashgate, 2009), Chap. 3.
58. See Samuel Taylor Coleridge, *Coleridge's Poetry and Prose*, ed. Nicholas Halmi, Paul Magnuson, and Raymond Modiano (New York: Norton, 2004), 68.
59. Coleridge, *Poetry and Prose*, 63.
60. Coleridge, *Poetry and Prose*, 62.
61. Bruno Latour, 'Love Your Monsters: Why We Must Care For Our Technologies As We Do Our Children,' *Breakthrough Journal* 2 (2012), accessed 8 May 2016, http://thebreakthrough.org/index.php/journal/past-issues/issue-2/love-your-monsters.
62. See 'An Ecomodernist Manifesto,' accessed 31 July 2017, http://www.ecomodernism.org/. Latour is not a signatory of the manifesto, although it clearly draws on some of his ideas.

63. James McKusick, *Green Writing: Romanticism and Ecology* (Basingstoke: Palgrave, 2000), 47.
64. For the poem's sources, see Jerome McGann's endnote in Byron, *CPW*, IV, 459–460; see also, R.J. Dingley, '"I had a Dream.....": Byron's "Darkness",' *Byron Journal* 9 (1981): 20–33; Catherine Redford, '"No love was left": The Failure of Christianity in Byron's "Darkness",' *Byron Journal* 43 (2015): 131–140; Jeffrey Vail, '"The Bright Sun Was Extinguish'd": The Bologna Prophecy and Byron's "Darkness",' *Wordsworth Circle* 28 (1997): 183–192. Buffon had noted in the *Natural History* (I, 79) that at some future point 'the sun itself will be extinguished.'
65. Brassier, *Nihil*, 223–230. Lyotard's essay is entitled 'Can Thought go on Without a Body?' and is published in *The Inhuman* (1991).
66. Brassier, *Nihil*, 223.
67. Brassier, *Nihil*, 224.
68. Brassier, *Nihil*, 230. Brassier's arguments can be connected to Morton's suggestion in *Hyperobjects* that 'the end of the world has already occurred' (7).
69. Buffon, *Natural History*, IX, 382.
70. Wood, *Tambora*, 69.
71. Wood makes the interesting suggestion that the Creature 'bears the mark of the famished and diseased European population by which [Mary Shelley] was surrounded by that dire Tambora summer' (*Tambora*, 66), although the worst part of the crisis would occur after the Shelleys had returned to England.
72. See Nigel Clark, 'Volatile Worlds, Vulnerable Bodies: Confronting Abrupt Climate Change,' *Theory, Culture & Society* 27, 2–3 (2010): 31–53.
73. Kate Rigby, *Dancing*, 79, 77. See also Olivia Murphy, 'Apocalypse Not Quite: Romanticism and the Post-human World,' in *Romanticism and Sustainability: Endurance and the Natural World, 1780–1830*, ed. Ben P. Robertson (Lexington Books: Lanham MD, 2015), 245–59.
74. For a valuable discussion of *Chillon* in the contexts of 1816 and Byron's more general interest in the 'last of the race' myth, see Fiona Stafford, *The Last of the Race: The Growth of a Myth from Milton to Darwin* (Oxford: Oxford University Press, 1994), Chap. 7.
75. Buffon, *Natural History*, IX, 396.
76. Buffon, *Natural History*, IX, 401. For a suggestive analysis of this aspect of Buffon's thought, see Noah Heringman, 'Deep Time at the Dawn of the Anthropocene,' *Representations* 129 (2015): 56–85.

77. It is also treated ironically by Percy in *A Vindication of Natural Diet*, which reads the myth of Prometheus as revealing the connection between the discovery of fire, meat-eating, and disease (*Prose Works*, 78–79).
78. Andrew Griffin, 'Fire and Ice in Frankenstein,' in *The Endurance of* Frankenstein: *Essays on Mary Shelley's Novel*, ed. George Levine and U.C. Knoepflmacher (Berkeley: University of California Press, 1979), 49–73.
79. Timothy Morton, 'Byron's Manfred and Ecocriticism,' in *Palgrave Advances in Byron Studies*, ed. Jane Stabler (London: Palgrave, 2007), 155–170 (156). In relation to theatricality, it is interesting to note that first version of 'Mont Blanc,' discovered in the Scrope Davies notebook, is entitled 'Scene–Pont Pelissier in the vale of Servox.' Percy also uses the word 'scene' to describe the area around Mont Blanc several times in the *History* (152, 155, 164).
80. Byron, *Journals*, V, 99.
81. Byron, *Journals*, V, 105.
82. Andrew Hubbell, 'Our Mix'd Essence': Manfred's Ecological Turn,' *Byron Journal* 42 (2014): 5–20 (12).
83. Meillassoux, *After Finitude*, 5. Useful discussions of Meillassoux and Percy Shelley can be found in Greg Ellermann, 'Speculative Romanticism,' *SubStance* 44 (2015): 154–174; Evan Gottlieb, *Romantic Realities: Speculative Realism and British Romanticism* (Edinburgh: Edinburgh University Press, 2016), Chap. 4; and Chris Washington, 'Romanticism and Speculative Realism,' *Literature Compass* 12 (2015): 448–460. See also Steven Shaviro, *The Universe of Things: On Speculative Realism* (Minneapolis: The University of Minnesota Press, 2014), 57–59.
84. Earl R. Wasserman, *Shelley: A Critical Reading* (Baltimore and London: The Johns Hopkins University Press, 1971), 146–147.
85. Meillassoux, *After Finitude*, 10–11.
86. Angela Leighton, *Shelley and the Sublime: An Interpretation of the Major Poems* (Cambridge: Cambridge University Press, 1984), 58–72. Classic epistemological readings of the poem include Frances Ferguson, 'Shelley's *Mont Blanc*: What the Mountain Said,' in *Romanticism and Language*, ed. Arden Reed (London: Methuen, 1984), 202–214; I. J. Kapstein, 'The Meaning of Shelley's "Mont Blanc",' *PMLA* 62 (1947): 1046–1060; Wasserman, *Shelley*, 222–238.
87. Cian Duffy, *Shelley and the Revolutionary Sublime* (Cambridge: Cambridge University Press, 2005), 6. The key study behind this approach is C. E. Pulos, *The Deep Truth: A Study of Shelley's Scepticism* (Lincoln: University of Nebraska Press, 1962).

88. Duffy, *Revolutionary Sublime*, 114. Gottlieb also questions Wasserman's claim about 'Shelley's belief in a universal mind' (*Romantic Realities*, 162).
89. Clark, *Inhuman Nature*, 95.
90. Clark, *Inhuman Nature*, 100–101.
91. Shelley, *Complete Poetry*, III, 512. 1816 was probably a particularly good summer for rainbows due to the wet conditions (Mary Shelley, *Journal*, 111). See also Mary Shelley's description of the waterfall at the Nant d'Arpenas as '*more like cloud than water, imitating a viel of the most exquisite woof*.' The idea of sculpture is clearly suggested by her observation of the water striking 'an enormous rock resembling' a 'colossal Egyptian statue of a female deity' and 'concealing the lower part of the statue' (*Journals*, 113–114).
92. O'Neill suggests that 'the syntactical incompleteness mirrors the sense of reverie, of floating intuitively induced in the poet': Shelley, *Complete Poetry*, III, 511.
93. Shelley, *Complete Poetry*, III, 80.
94. *The Poems of Shelley Volume 1: 1804–1817*, ed. Geoffrey Matthews and Kelvin Everest (London: Routledge, 1989), 538.
95. Shelley, *Poems of Shelley*, 538.
96. Wasserman, *Shelley*, 227.
97. Morton, *Hyperobjects*, 15.
98. Morton, *Hyperobjects*, 1. In making this connection, I differ from Gottlieb who sees the mountain as 'clearly localized' and therefore not a hyperobject in Morton's sense (*Romantic Realities*, 169).
99. Morton, *Hyperobjects*, 195–196.
100. In a passing reference to Shelley's poem, Morton notes that 'the mountain comes in and out of phase,' another property that he associates with hyperobjects (*Hyperobjects*, 72).
101. Meillasoux, *After Finitude*, 112–113. For Meillasoux, crucially, this problem of diachronicity is qualitatively different from how correlationism makes sense of spatial distance because it does not deal with an event that is unwitnessed but with one that precedes or succeeds the possibility of witnessing (i.e. the very being of the universe in correlationist terms).
102. Wasserman, *Shelley*, 228.
103. Ferguson, 'What the Mountain Said,' 208.
104. Byron, *Letters and Journals*, V, 99.
105. 'Eternal' is an important term in other 1816 texts. Byron describes 'summits of eternal snow' in the 'Alpine Journal.' In Canto III of *Childe Harold's Pilgrimage*, the Alps are (using the language of human dwelling that we also find in 'Mont Blanc') 'The palaces of Nature, whose

vast walls /Have pinnacled in clouds their snowy scalps, /And throned Eternity in icy halls/Of cold sublimity' (Byron, *Complete Poetical Works*, II, 100, ll. 591–595). In the third letter to Peacock in the History, Shelley describes how the bases of the mountains in 'the valley of the Arve' are 'still covered with the eternal forests' (Shelley and Shelley, *History*, 147).

106. See also the description of an Alpine storm in Canto III of *Childe Harold's Pilgrimage*, which describes the 'mountain-mirth' of the 'loud hills,' 'as if they did rejoice o'er a young earthquake's birth' (Byron, *Complete Poetical Works*, II, 111, ll. 876–877).
107. Ferguson, 'What the Mountain Said,' 211.
108. Louise Economides, '"Mont Blanc" and the Sublimity of Materiality,' *Cultural Critique* 61 (2005): 87–114 (106).
109. Leask, 'Mysterious Voices', 187.
110. Leask, 'Mysterious Voices', 188.
111. Leighton, *Shelley and the Sublime*, 58–61; Duffy, *Revolutionary Sublime*, 111–113.
112. Duffy, *Revolutionary Sublime*, 119–120.
113. Shelley, *Complete Poetry*, III, 75, 77; see Duffy, *Revolutionary Sublime*, 105.
114. Byron, *Journals*, V, 97.
115. At this point in the poem, the mountain is much more than a topographical feature. It embodies 'Power,' which Wasserman (along with other critics) plausibly reads as causation or Necessity (*Shelley*, 229–231).
116. 'daedal, adj.' *OED Online*. March 2017. Oxford University Press. http://0-www.oed.com.wam.leeds.ac.uk/view/Entry/46834?rskey=kQb2Uk&result=2 (accessed May 20, 2017).
117. Duffy, *Revolutionary Sublime*, 125.
118. Duffy, *Revolutionary Sublime*, 124.
119. Meillassoux, *After Finitude*, 20.
120. Ferguson, 'What the Mountain Said, 210.
121. Ferguson, 'What the Mountain Said', 213.
122. Robinson, *Snake*, 125.
123. Wasserman, *Shelley*, 238.
124. Kapstein, 'Shelley's "Mont Blanc"', 1057–1058.
125. Duffy, *Revolutionary Sublime*, 119.
126. Economides, 'Sublimity of Materiality', 108–109.
127. Colebrook, *Death*, 23.
128. After writing my account of 'Mont Blanc,' I came across Greg Ellermann's discussion of its ending in relation to Meillassoux's ideas. Ellermann also seeks to move beyond epistemological readings by

understanding 'Power' as an absolute contingency that exists beyond the correlation and which can be intuited through the contingency of language: 'by insisting on their own status as mere words and letters—on their own fundamental meaninglessness, that is—the lines and marks that comprise "Mont Blanc" bring into view the contingency of all things' ('Speculative Romanticism,' 169). While this analysis is astute and suggestive, I find no such insistence in the poem.

129. Clark, *Inhuman Nature*, 50.
130. Meillassoux, *After Finitude*, 115. For the 'withdrawnness' of objects, see Morton, *Hyperobjects.*
131. Bronislaw Szerszynski, 'Reading and Writing the Weather: Climate Technics and the Moment of Responsibility,' *Theory, Culture and Society* 27 (2010): 9–30 (25).

CHAPTER 4

The 'Year Without a Summer' and the Politics of Climate Change

Abstract This final chapter focuses on periodical writings about the 'distresses' of the Regency, revealing the interplay of the 'natural' and the 'cultural' that makes an environmental catastrophe. It shows how reformers like William Cobbett sought to argue that poverty and famine, rather than part of the order of nature, were politically produced. However, through a reading of the *Examiner's* republication of Coleridge's poem 'Fire, Famine, and Slaughter', it also reveals the emergence of a more complex idea of agency in which environmental and political factors are seen to work in tandem. This analysis provides a platform for some final considerations about the difficulty of assigning agency in relation to present-day climate change, and the ways in which the Anthropocene both aggrandises and humiliates humanity.

Keywords William cobbett · Samuel taylor coleridge · Leigh hunt
New materialism · Regency · Anthropocene

This book has focused on two case studies: the compilation of eyewitness accounts of the 1815 Tambora eruption and its aftermath put together by the Raffles administration in Java, and the 1816 writings of Byron and the Shelleys. It has analysed Tambora as a material-discursive assemblage and shown how the 'Year without a Summer' contributed to the Diodati Circle's reflections on the precarity of human life on a changeable and unpredictable planet. It has not only examined how a large-scale

D. Higgins, *British Romanticism, Climate Change, and the Anthropocene*,
DOI 10.1007/978-3-319-67894-8_4

catastrophe threatens anthropocentric ideas of agency, but also how it is mediated through textual assemblages rather than the work of individual authors. Tambora's other discursive traces in British writing of the period tend to be more fragmentary and oblique. For example, the eruption and its aftermath inform late Romantic texts about polar exploration because, although the eruption led to cooler temperatures across much of the world, it also caused warming and ice-melt in the Arctic.[1] Thus we find Robert Southey writing to Herbert Hill with some excitement at the end of 1817: 'Have you heard the great news from the North? Some great convulsion seems to have broken up the ice about the pole.'[2] The February 1818 number of the *Quarterly Review* included an influential article on the Arctic by Sir John Barrow, a prolific reviewer and the Second Secretary to the Admiralty. Barrow, who would go on to support a number of important polar expeditions, argued that the warming polar climate offered a great opportunity for exploration, particularly with the aim of discovering the Northwest Passage. As Adeline Johns-Putra has shown, Barrow's article and its concern with climate change offers an important context for an eco-historical reading of Eleanor Anne Porden's poem *The Arctic Expeditions* (1819).[3] Other critics have focused on representations of weather within Britain and particularly the work of John Keats. Jonathan Bate's ecological reading of 'To Autumn' (1820) contextualises the poem in relation to the Tambora crisis by pointing out that the weeks in August and September 1819 preceding its composition were considerably sunnier and warmer than their equivalents during the preceding three years.[4] And Alexandra Harris suggests that Keats's sonnet 'After dark vapours have oppressed our plains' (1817) responds to the 'aberrant weather' by showing that 'the atmosphere itself is ill.'[5] Astute as these readings are, it is clear that Tambora does not provoke a significant body of catastrophe literature by canonical writers working within Britain, for the obvious reasons that they were largely protected from its effects and that it was not widely reported.

However, this concluding chapter identifies a set of texts that have not previously been analysed in relation to the Tambora crisis: writing about politics and weather in British periodicals of 1816–1817. As with my other case studies, I understand these writings not so much as individual utterances, but as forming an intertextual assemblage that is an insertion into the crisis rather than merely a commentary on it. They focus particularly on the question of whether the travails of the poor are largely the result of a corrupt political system or of the bad weather. My analysis will show that they therefore speak to one of the most difficult issues of the Anthropocene: the relationship between power, inequality, and climate change. At the same time, I am also attentive to their particular historical valencies. They should be read as part of a broader public discourse about the so-called 'distresses' affecting Britain after the end of the Napoleonic Wars, which were caused by a complex matrix of factors: high taxation; the post-war collapse of the coal, iron, and textile industries; agricultural foreclosures; unemployment; and the climatic conditions.[6] England, Scotland, and Wales do not seem to have experienced a famine per se—Ireland is a different matter—and their subsistence crises were probably not as severe as those affecting Continental nations.[7] Nonetheless, it is clear that for some parts of the population, the very high food prices, the depressed post-war economy, and the dreadful harvests of 1816 and 1817 exacerbated an already perilous situation of subsistence living and caused serious hardship. The result was intense political debate and public unrest: there were agrarian riots across the country in May 1816 due to the price of food.[8] In July, 'a group of starving colliers from [...] Staffordshire achieved national notice by yoking themselves like beasts of burden to several huge wagons laden with coal and petitions for parliamentary reform, which they began to haul along the main highways to London as symbolic gifts to the Regent himself.'[9] In August, the *Edinburgh Review* claimed that 'at no former period of the history of this country, was so great and general a distress known to prevail, as that which has lately visited us, and of which the pressure still unhappily continues.'[10] By November, even the sober *Farmer's Magazine* noted that 'the prospects of the ensuing year are of a very unfavourable nature. Indeed they present a more gloomy aspect to the labouring class of the community than at any period during the last fifty years.'[11] The same month saw a large reformist meeting at Spa Fields that had the aim of 'petitioning the Prince Regent upon the present distressed state of the country' and at which the radical leader Henry Hunt railed against government corruption and high taxation.[12] A second meeting on 2 December, called after Hunt had been refused access to the Regent, ended in disorder when a group of protesters, possibly incited by a government agent, moved towards the Tower of London and looted a gun shop before being stopped by soldiers. This event contributed to the passing of the Seditious Meetings Act and the suspension of Habeas Corpus early in 1817.

Central to the febrile political discourse of the period was William Cobbett. His *Political Register* had long been a thorn in the side of the government and in November 1816 he launched a hugely successful mass-market version priced at only 2d, much to the consternation of conservatives.[13] (As a result of the government clampdown, he fled to America in early 1817, where he remained until October 1819.) Writing at the end of August 1816, Cobbett notes that the weather had improved to the extent that it was likely to 'make up for the effects of the wet, [so] as to give us a good average crop of wheat.' (This turned out not to be the case.) He hopes that, as a result, the nation will 'escape that worst of all calamities, riots and violences, arising from a high price of food.' After this disarmingly bland opening, Cobbett returns to his usual pointedness. What would be truly 'calamitous,' he suggests, would be if the bad weather were to continue, for 'it would be manifest [...] that the government could not help bad seasons; and it is quite certain that their supporters would impute all the miseries of the country to the rain and the cold.'[14] Cobbett, like many other reformers, understands 'the national misery' not as caused by 'the dispensations of Providence' but by the 'intolerable load of taxes.'[15] He would go on to make this point forcefully two months later in 'To the Journeymen and Labourers of England, Wales, Scotland, and Ireland,' the article that launched the mass-circulation 'Two-Penny Trash' and which sought to address 'a nascent and as yet unexploited national audience.'[16] Cobbett emphasises that the source of the power of Britain, 'so poor in climate and soil compared with many others,' is the '*labour* of its people'. Without them, and despite its impressive infrastructure and manufacturing goods, 'the country would be a wilderness.'[17] However, the cause of the 'miseries' of the labouring classes is not a recalcitrant environment, but '*the enormous amount of the taxes*.'[18] He notes that corn is currently '*high in price*' due to the '*badness of the crop*' in order to emphasise that farmers are also victims of the confluence of environmental, economic, and political factors.[19] The implication is that a parliamentary reform, and subsequent reduction of the tax burden on the poor, would make them much less vulnerable to climatic contingencies.

It was important for reformers to emphasise that the key causes of dearth were mainly political rather than environmental processes in order to counteract conservative arguments—such as those of Cobbett's hated 'Parson Malthus'—that saw the suffering of the labouring classes as an inevitable part of the order of nature and therefore not worth attempting to address in any systematic way. This quietistic view is clearly articulated by the conservative *Courier* newspaper, in an August 1816 article to which Cobbett may well have been responding:

> It is a wretched delusion attempted upon the people to tell them that Parliament can furnish a remedy for their distresses. It is recommending them to make no effort, to take no thought for themselves, but to give credit to such a monstrous assertion, that Parliament can change the course of nature and of the seasons—convert, by a simple resolution, bad weather into good, wet into dry, and work, like JOSHUA, a miracle with the sun—not commanding it indeed to stand still, but to shine.[20]

To believe that governmental action can do anything to ameliorate the effects of natural phenomena such as the weather and seasonal variation is not only delusional and prevents individuals from taking responsibility for their own lives, but is a kind of 'monstrous' blasphemy that seeks to arrogate divine power to human institutions. (It is disheartening, to say the least, that this sort of disingenuous fatalism is so common in present-day discussions of climate change and related catastrophes.) Although the *Courier* generally accepted that Britain, and Europe more widely, was in a distressed state, it consistently refused to accept that the situation could be improved by structural change or to show understanding of the desperation of the poor. Thus, a summary of Continental papers in July 1816 notes 'the dreadful ravages of storm and tempest, of lightning, thunder, and rain' leading to widespread flooding, the destruction of food supplies, and despair and poverty. And yet the same summary makes no connection between these terrible conditions and the 'increasing number of suicides on every part of the Continent,' blaming them instead on the French Revolution and its associated philosophical 'heroes of impiety and atheism.'[21] Further Continental reports of the severe 'damage done by bad weather' appear in August 1816, 'accompanied by judicial proceedings in the different Courts throughout France, which show the dreadful deterioration of morals produced by the revolution.'[22] Once again, no connection is made between environmental catastrophe and individual transgressions, which the *Courier* blames entirely on pernicious radical ideas.

The reformist counter argument about the politics of dearth is made economically by a short poem in the radical periodical *The Black Dwarf* entitled 'Want, Famine, and Mortality,' published in May 1817.

It suggests that the 'health' and 'contentment' of a once industrious and rich nation in which 'industry's plenteous store was seen' has been replaced by 'gen'ral gloom,' for 'famine's reign the land pervades.' The final stanza pointedly asks about its causes:

> What adverse fate destroys our life?
> Are nature's laws with man at strife?
> Or is it man's delirious sway,
> Destroys our wealth, and wastes our lives away?[23]

For reformers, the answer was obvious: the distresses were not caused by natural variability (the 'adverse fate' of bad weather), but due to the 'delirious sway'—a kind of diseased madness—of a corrupt political and economic system. As a correspondent in the *Examiner* asked, alluding to recently imposed restrictions on the import of grain that contributed to the high prices, 'have not the blights and rains had a terrible ally in this Corn Bill?'[24] Against Malthus and other conservatives, reformers did not think that famine was built into 'nature's laws.'[25] Therefore the 'strife' was not between humanity and elemental forces, but between the poor and rich. Writing in 1817, towards the end of another wet summer, Leigh Hunt emphasised this point by suggesting—perhaps rather too ingeniously to be convincing—that 'our climate alone is a host against the meditators of arbitrary government,' for the discomforts of the nation's weather meant that the English needed '*freedom*' to make life tolerable whereas the inhabitants in warmer Continental countries were more tolerant of oppression.[26] For Hunt, therefore, the 'vapours, and clouds, and storms' described in Thomson's *Seasons* were potentially on the side of Reform.[27]

The idea that inequality was the result of human constructions rather than natural forces was also suggested by the *Examiner*'s publication of Percy Shelley's 'Hymn to Intellectual Beauty' in January 1817. As I discussed in Chap. 3, the poem understands the apprehension of the sublime as offering an antidote to the 'frail spells' of religion and the despotic political systems that Shelley believed that religion supported.[28] The poem's aspiration that such an apprehension 'wouldst free / This world from its dark slavery' and lead to the love of 'all human kind' resonates within a periodical that consistently connected a highly aestheticized literary sociality with political reform.[29] But the poem's radicalism—its rejection of a providential understanding of the universe in favour of a catastrophic necessitarian power—also exists uneasily alongside the *Examiner*'s more moderate ideas. Directly after the poem, the paper prints an article entitled 'Reform' that includes extracts from a speech by the Whig politician and lawyer John Philpot Curran attacking British foreign policy in relation to the American and French Revolutions, and the treatment of Ireland.[30] Ignoring the 'admonition of Providence,' he argues, Britain has created wars through its own excessive militarism.[31] Curran paints a remarkable picture of a nation 'fall[ing] to pieces'; a meteor rapidly descending through the air: 'like other falling bodies, it descends with a constantly increasing velocity of precipitation! How far are we from the earth? How long will it take to pass the space which remains?' The speech blames the country's distresses not on climatic contingency but on a bloated political state apparatus. The result is a dangerous 'infection,' for when a man is no longer able to feed his family, 'the social tie which binds him is broken, and famine becomes the recruiting officer for revolution.' Curran calls for a prophet: a 'John the Baptist' who would be able to make sense of the 'exasperation' felt by 'English minds' at the 'famine and oppression' that beset them: 'You feel what your grievance is—but what is the remedy? Neither patient nor doctor know our evils, we feel impending, famine and the dread of a military despotism.' This sense of crisis and desire for apocalyptic revelation is not too far from Shelley's poetry. And yet Curran's speech does not in fact call for the kind of catastrophic overturning that the reference to John the Baptist implies. He supports an enlargement of the franchise, but certainly not universal suffrage, and is keen to distinguish his group of gentlemanly reformers from more populist movements, so as to show that 'Parliamentary Reform does not consist in breaking windows, or getting drunk in the streets.' Read alongside Curran's speech, the 'Hymn' may well have seemed abstract and aestheticized, and yet its presentation of a universe without Providence was in some respects far more radical.

The use of the word 'famine' by Curran and in *The Black Dwarf* was fairly unusual in the context of reformist periodical discourse. The more general term 'distresses' was much more prevalent, with its sense not only of adversity but of the country under pressure and constraint from poor governance, high taxes, and corruption. (I refer to the word's connection to 'distrain' and their shared root in the Latin *districtia*, meaning to squeeze or draw tight.) Whereas 'distresses' was a term of established political debate, including within Parliament, 'famine' had more revolutionary and apocalyptic overtones, as is apparent in the November 1816 republication of Coleridge's poem 'Fire, Famine, and Slaughter' (1798) in the *Examiner*. The poem describes itself as a 'war eclogue,' and maintains the dialogic style of the classical eclogue while moving it from pastoral to anti-pastoral. It 'recounts a meeting between the three personifications named in its title, which are seemingly based on the three witches in *Macbeth*. All three, it transpires, had been sent there to do their terrible work of destruction by a man whose name is so terrible, so unspeakable, they dare not pronounce it'.[32] The poem is in part an attack on Pitt—'Letters four do form his name'—but this attack is based more broadly on a depiction of British foreign policy as cruel and destructive.[33] When Coleridge came to republish the poem himself in *Sibylline Leaves* in July 1817, he added an 'apologetic preface' where he suggests that the poem shows no animus against the actual Pitt. As John Barrell puts it, Coleridge argues that there is 'a kind of joyousness involved in the invention and development of figurative utterances that is entirely at variance with the intensity of implacable hatred.'[34] Coleridge's disavowal of the poem's seriousness and the preface's concomitant attempt to split poetry from politics was, I suspect, in part prompted by the *Examiner*'s co-option of the text a few months earlier.

The republication of the poem continued the *Examiner*'s frequent attacks on the Lake Poets as political turncoats; it also spoke to the political instability and national crisis affecting Britain in late 1816. In the leading article to the edition that contained Coleridge's poem, the newspaper criticised Henry Hunt as a dangerous demagogue. Leigh Hunt attempts to distinguish the sensible reformism of the *Examiner* and Cobbett from 'interference of mere noise and violence,' complaining about 'men who have talked at public dinners about heads gracing lamp-posts, and who disturb the solemn inflexibility of Reform by doing their best to identify it with Revolution.'[35] The *Examiner* often presented itself as offering a middle ground between extremists on both sides, who supposedly confused sensible reform with violent revolution. However, this did not prevent the paper from raising the spectre of revolution as a necessary consequence of grotesque inequality exacerbated by bad weather. In a three-part article in August and September on the 'Relief of the National Distresses,' it presents a nation in crisis while 'the Sinecurist [...] sits at ease amidst his luxuries': 'in the mean time, the rains come down, the distresses increase, bankruptcy crowds upon bankruptcy,—the poor are compelled to retrench, some of them almost to starve,—the fields and roads are swarming with beggars.'[36] The article simultaneously mocks the reactionary 'cant about Revolutions' in conservative organs such as the *Courier* newspaper, and warns of the consequences of a lack of reform: 'people in general no more wish for Revolutions, than they do for bankruptcies, oppression, and starvation; but if the latter come, and are not properly looked after, then they may wish for a Revolution as the natural crisis of the disorder, and their only chance for restored health.'[37] In Coleridge's poem, this idea of revolution as perhaps the only way to cure the sickness of the body politic finds an analogue in Famine's cry that:

> Wisdom comes with lack of food.
> I'll gnaw, I'll gnaw the multitude,
> Till the cup of rage o'erbrim:
> They shall seize him and his brood.

Just as 'Darkness' describes a warring world in which 'the pang /Of famine fed upon all entrails,' here Famine 'gnaws' on the bodies of the people until they become violent. Reprinting 'Fire, Famine, and Slaughter' allowed the *Examiner* to present revolution not as part of reasoned and nuanced political discussion—in which it needed simultaneously to be threatened and disavowed—but through prophecy as a spell-like performative utterance.

The poem also allowed the paper to vividly dramatise the sufferings of the poor, in a passage that must have seemed particularly stinging after the national self-congratulation of 1815:

> FAMINE
> Thanks, Sister! Thanks! the men have bled,
> Their wives and children faint for bread.
> I stood in a swampy field of battle
> With Bones and Skulls I made a rattle
> To frighten the wolf and the carrion crow
> And the homeless dog but they would not go
> So off I flew, for how could I bear
> To see them gorge their dainty fare,
> I heard a groan and a peevish squall
> And thro' the chink of a cottage wall
> Can you guess what I saw there?

BOTH
Whisper it, Sister! In our ear.

FAMINE
A baby beat its dying mother
I had starv'd the one, and was starving the other!

This emphasis on the domestic and familial consequences of warfare for those left at home was of course a common reformist trope throughout the Romantic period. The poem's vividness and willingness to allow its readers to witness the grotesque suffering of the poor through 'the chink of a cottage wall' is quite different from the *Examiner*'s usual more intellectualised rhetoric; more akin, in fact, to the emotive writings of Cobbett. Coleridge's emphasis on the destruction wreaked on the rural poor by government policy was also useful to the *Examiner* as a way of defending the popular clamour for reform. In a two-part article published about three weeks before the poem, Hunt complains about conservative attacks on 'inflammatory' speeches and lists various Establishment figures as 'great inflamers.' What is really inflammatory and causes 'unlawful acts of violence,' he argues, is a corrupt system that ensures that 'a single individual is not to bear the diminution of a sinecure of twenty-four thousand a year, but to heap up riches and increase his land, while his fellow creatures are dropping with grief and hunger round about it [...] These are the things that make the popular indignation blaze up.'[38] Similarly, in the poem, the true 'inflamers' are not the Irish rebels, but the British troops and their leaders: 'By the light of his own blazing Cot / Was many a naked rebel shot.' Here again, the poem focuses on the destruction of a rural cottage as a metonym for state oppression.

Two weeks after publishing Coleridge's poem, and one week after the Spa Fields riot, the *Examiner* published a leading article on 'Disturbances in the Metropolis.' Those reactionaries (such as writers for the *Times* and the *Courier*) who wring their hands about rioting in London, Hunt notes, 'will advocate with roaring smiles the wars and slaughters, the infinite deaths and miseries, that take place in behalf of Royalty.' He presents some of these 'criers out against riot' as formerly violent Jacobins, mentioning the republication of Coleridge's poem, a number of Southey's poems, and the coterie's 'violent revolutionary speeches.' The violent intensity of Coleridge's poem typifies not only the excessive zeal of these individuals, but also their hypocrisy: 'these advocates of Fire, Slaughter, and Famine, whenever such furies are on their side of the question, now pretend to groan in spirit, and denounce the excesses of one day's rioting!'[39] Republishing Coleridge's poem allowed the *Examiner* to depict the distresses of the Regency using more heightened and emotional terminology than was normally possible within its reasonable reformist discourse: apocalyptic rather than realistic; hyperbolic rather than nuanced; grotesque rather than ironic. It allowed the paper to show (rather than argue), in a move that would be repeated by Hunt's friend Percy Shelley in the 'Masque of Anarchy' (composed 1819), that the 'inflamers' were not the people attending reformist meetings, or even those participating in public disorder, but those who sought to preserve a corrupt *status quo*. The poem's complex prosopopoeia was mobilised around the question of 'who speaks for the people,' and its dependence on an unspeakable word could hardly fail to resonate at a time when freedom of speech was about to be severely curtailed.

I have spent some time addressing the republication of Coleridge's poem not only because it sheds light on reformist discourse on the Regency crisis, but also because it intersects with this study's concern with human and natural agency and the politics of climate change in the Anthropocene. The three sisters that the poem brings together not only personify different phenomena, but different sorts of phenomena: an elemental force; the extreme scarcity of food; and deaths in battle. The three figures present themselves as emissaries of Pitt and therefore might seem to be straightforwardly under human control. But the mysterious phrase, 'four letters form his name,' suggest that Pitt may be more than just a person; hence Barrell's argument that the four letters could also refer to 'King.'[40] In the context of the *Examiner*'s writings of 1816, 'Fire, Slaughter, and Famine' can be understood as aspects of a corrupt and unequal socio-political system intersecting with a global environmental crisis, rather than the servants of a single demonic individual. They constitute a kind of grim assemblage of nonhuman and human agencies. Of course, reformist writers were keen to assign agency and responsibility as much as possible to powerful people in order that the situation might be remedied. But there was also an inchoate understanding of the role of other factors in the distresses afflicting the country—perhaps even a notion of distributed agency—apparent, for example, in the suggestion that 'the blights and rains had a terrible ally' in the Corn Bill. Although writers of the period did not use the modern language of risk, they understood the poor weather as a natural hazard which had particularly deleterious effects due to the vulnerability of parts of the British population: a vulnerability in reality caused by a range of political, economic, environmental, and social factors, but which reformers found expedient to blame on sinecurism and taxation.

There is clearly a political and ethical value in holding humans and their institutions, rather than elements or natural objects, principally responsible for the effects of climate change: hence the focus of modern environmentalism on government policy and the fossil fuel industry. The alternative might seem to be a kind of fatalism, or a Malthusian belief in the naturalness of exploitation and inequality. But the Anthropocene complicates 'moral and political accounting.'[41] First, global warming is not simply the result of human activity, but rather a complex assemblage of matter and energy flows in which we play an important role. As Latour puts it, in the Anthropocene we are experiencing an 'utter confusion between subjects and objects.'[42] And secondly, although the complicity of powerful vested interests in maintaining carbon capitalism is obvious, we all contribute to the problem through our daily lives: eating, travelling, heating and cooling our homes, having children, publishing academic books. To that extent, climate change requires 'species thinking' even as it also demands close attention to the reality that 50% of carbon emissions are caused by the richest 10% of people in the world, a group that is likely to include most readers of this study.[43] As has often been stated, climate change produces, and is generated by, brutal inequalities. Those communities that have contributed the least to it are often the most vulnerable to its effects. There may be an instructive parallel between the aestheticised view of the climate crisis taken by sheltered European elites in 1815–1818, and the complacent attitude towards present-day global warming taken by many in the West. For Raffles, the Tambora eruption was extraordinary, but could be controlled and mediated through the apparatus of colonialism. Similarly, the Diodati Circle's complex accounts of human precarity were contingent on Byron and the Shelleys being themselves relatively sheltered from the climate crisis. I have therefore tried to take a critical distance from my case studies' tendency to heroise the exceptional individual's sublime encounter with the elements and to contextualise them in relation to the vulnerability of less privileged groups. Then as now, such groups were not well served by canonical forms of representation and mediation. If the new materialism and material ecocriticism are going to help us to understand the Anthropocene, they need to find ways to analyse assemblages that shed light on the political, social, cultural, and economic inequalities associated with climate change rather than flattening them through a model of agency that does not attend sufficiently to how power is distributed. It is easy to make vague assertions about material-discursive entanglements; much harder to unpack and theorise the different valencies and weightings of those entanglements. It is also relatively easy to write aerily about catastrophic assemblages from the comfort of a centrally heated office in a relatively temperate climate. This book has attempted to make the case for a historicist criticism that is sensitive to the interactions of matter, discourse, and power. However worthwhile such an approach, no work of criticism can do justice to the devastating effects of the Tambora eruption and the suffering and death that it caused. To that extent, my analysis cannot help but have rehearsed the apparently disinterested, imperialising mode of Raffles's 'Narrative.'

Addressing the relationship between morality and the agency of assemblages, Bennett suggests that an individual's 'ethical responsibility' may now reside 'in one's response to the assemblages in which one finds oneself participating: Do I attempt to extricate myself from assemblages whose trajectory is likely to do harm? Do I enter into the proximity of assemblages whose conglomerate effectivity tends towards the enactment of nobler ends?'[44] However, there are surely assemblages from which an individual will be unable to extract herself, short of self-annihilation, and which she therefore may wish to influence from within. As Ian Buchanan puts it, 'if the ethical task is to respond to the assemblage that we are participating in, then that surely means we have to start with the assemblage we are all participating in, namely the planet as a habitable place.'[45] The texts that I have addressed in this study do not offer straightforward ethical formulae for dwelling with the Anthropocene. But they do have much of interest to say about the precarity of life on Earth, and the imbrication of human and elemental forces that come together to form an environmental catastrophe. If the new materialists emphasise the interconnection between different sorts of vital energies, the more nihilistic branch of speculative realism shows human beings as profoundly disconnected from uncaring objects. Both strands of thought are evident in

Romantic writing, and both are significant to our current predicament. I hope that I have shown how these apparently abstract philosophical ideas can speak to the politics of climate change.

In *The Great Derangement*, Amitav Ghosh argues that literary fiction has found it difficult to engage with the topic of climate change because of the emphasis on order, probability, and human agency that the novel developed in the nineteenth century. He suggests, too, that the essence of the Anthropocene 'consists of phenomena that were long ago expelled from the territory of the novel – forces of unthinkable magnitude that create unbearably intimate connections over vast gaps in time and space.'[46] One might well take issue with the sharp division that Ghosh makes between literary and genre fiction, but he nonetheless has a point. Coming as it does before the full development of nineteenth-century realism, Romantic writing speaks powerfully to the sublime, catastrophic, and discontinuous entanglements of our world. As we have seen, its apparent anthropocentrism is meliorated by its sense of the power of nonhuman forces and its merging of political and natural catastrophes. By focusing on textual assemblages rather than individual authors, I have tried to suggest a critical method that is not only appropriate for my case studies, but also for addressing the challenge that climate change presents to the bourgeois valorisation of individual agency. *Writing Tambora* has addressed the idea of the catastrophe as an exceptional event: a huge volcanic eruption or even the death of the sun. It has also been concerned with how imagining such events can have a catastrophic effect on our understanding of the human relationship to the universe, by making us aware of the array of unpredictable forces with which we interact. But a more quotidian sense of the catastrophic will become prevalent as we try to adapt—practically, politically, and epistemologically—to climate change as it intertwines with our mundane experiences and activities. The environmental humanities are rightly concerned with addressing the present-day ecological problems to which we contribute even as they beset us and the other entities with whom we share the planet. Despite the remarkable efforts of many individuals and institutions, these problems may not be soluble. We can learn much from the Romantic period about how the Anthropocene simultaneously aggrandises and humiliates humanity.

Notes

1. Gillen D'Arcy Wood, *Tambora: The Eruption that Changed the World* (Princeton and Oxford: Princeton University Press, 2014), Chap. 6.
2. Robert Southey, letter to Herbert Hill, 23 December 1817, in *The Collected Letters of Robert Southey: A Romantic Circles Electronic Edition: Part Five: 1816–1818*, ed. Tim Fulford, Ian Packer, and Lynda Pratt, http://www.rc.umd.edu/editions/southey_letters/Part_Five/index.html, paragraph 3. Interestingly, Southey thought that the activity of marine volcanoes was the likely cause. He mentions the warming Arctic in several other letters of late 1817 and early 1818.
3. Adeline Johns-Putra, 'Historicizing the Networks of Ecology and Culture: Eleanor Anne Porden and Nineteenth-Century Climate Change,' *Interdisciplinary Studies in Literature and Environment* 22 (2015): 27–46.
4. Jonathan Bate, *The Song of the Earth* (London: Picador, 2000), 102.
5. Alexandra Harris, *Weatherland: Writers & Artists Under English Skies* (London: Thames & Hudson, 2015), 258.
6. For the crisis of this period, see John D. Post, *The Last Great Subsistence Crisis of the Western World* (Baltimore and London: The Johns Hopkins University Press, 1977) and R.J. White, *Waterloo to Peterloo* (Harmondsworth: Penguin, 1957).
7. For variations in the crisis across Europe, see Post, *Subsistence*, Chap. 2. For the Irish famine of the late 1810s, see Wood, *Tambora*, Chap. 8.
8. Post, *Subsistence*, 70.
9. Robert Lapp, *Contest for Cultural Authority: Hazlitt, Coleridge, and the Distresses of the Regency* (Detroit: Wayne State University Press, 1999), 52. For more detail on the riots in East Anglia, see Kent Linthicum, '"Bread or Blood": Climate Insecurity in East Anglia in 1816,' accessed 2 June 2017, http://www.historicalclimatology.com/blog/bread-or-blood-climate-insecurity-in-east-anglia-in-1816.
10. 'Art. I. The Speech of Charles C. Western,' *Edinburgh Review*, June 1816, 255. (This number of the *Edinburgh* was actually published in August.)
11. 'T.', 'On the Deficiency of the present Crop, and Economy recommended in its Consumption,' *The Farmer's Magazine*, November 1816, 463.
12. 'Meeting at Spa-Fields,' *The Examiner*, 17 November 1816, 730.
13. Richard Ingrams, *The Life and Adventures of William Cobbett* (London: Harper Perennial, 2006), 124–129.
14. William Cobbett, 'Corn Crops,' *Political Register*, 31 August 1816, 286.
15. Cobbett, 'Corn Crops,' 286.
16. Leonora Nattrass, *William Cobbett: The Politics of Style* (Cambridge: Cambridge University Press, 1995), 109.

17. William Cobbett, 'To the Journeymen and Labourers of England, Wales, Scotland, and Ireland,' *Political Register*, 2 November 1816, 545–546.
18. Cobbett, 'Journeymen,' 547.
19. Cobbett, 'Journeymen,' 570–571.
20. *The Courier*, 13 August 2016, 3. A couple of weeks previously, the paper had asked 'what could the Parliament do? Can it control the seasons—can it create employment?' (1 August, 2).
21. *The Courier*, 23 July 1816, 2.
22. *The Courier*, 7 August 1816, 2.
23. 'G.K.', 'Want, Famine, and Mortality,' *The Black Dwarf*, 21 May 1817, 271.
24. 'The Corn Bill,' *The Examiner*, 8 December 1816, 710.
25. For a brilliant attack by Cobbett on the Malthusian 'law of nature,' see 'To Parson Malthus,' *Political Register*, 8 May 1819, 1029–1048.
26. [Leigh Hunt], 'On the Necessity of being Watchful during the Vacation of Parliament,' *The Examiner*, 10 August 2017, 498.
27. Hunt, 'Necessity,' 498.
28. For a full analysis of the poem along these lines, see Cian Duffy, *Shelley and the Revolutionary Sublime* (Cambridge: Cambridge University Press, 2005), 97–106.
29. 'Original Poetry,' *The Examiner*, 19 January 1817, 41.
30. The juxtaposition of the 'Hymn' and the article on 'Reform' is noted by Nicholas Roe in *Fiery Heart: The First Life of Leigh Hunt* (London: Pimlico, 2005), 288. Roe also notes that the *Examiner* published Keats's sonnet 'After dark vapours have oppressed our plains' around this time and that the poem might therefore be read as articulating political hopefulness (294).
31. All quotations from the speech are from 'Reform,' *The Examiner*, 19 January 1817, 42.
32. John Barrell, '"Fire, Famine, and Slaughter",' *Huntingdon Library Quarterly* 63 (2000): 276–298 (286).
33. I refer throughout to the version of the poem published in 'Mr Coleridge and the Edinburgh Reviewers,' *The Examiner*, 24 November 1816, 744. The first of the two immediate contexts for the poem in 1798 was the Royalist insurgency in the La Vendée in 1793 and 1796, which was encouraged and supported by Pitt's ministry, but did not receive British reinforcements and was ultimately brutally swept aside by republican forces. The second was General Lake's repressive regime in Ulster in 1797: 'Sisters!', cries Fire, 'I from Ireland came! / Hedge and corn-fields all on flame.'
34. Barrell, 'Fire,' 289.
35. [Leigh Hunt], 'Mr. Hunt of Bristol and the Editor of the Times,' *The Examiner*, 24 November 1816, 738.
36. [Leigh Hunt], 'Relief of the National Distresses,' *The Examiner*, 11 August 1816, 497.

37. [Leigh Hunt], 'Relief of the National Distresses,' *The Examiner*, 1 September 1816, 545.
38. [Leigh Hunt], 'Who are the Inflammatory?', *The Examiner*, 27 October 1816, 673. The 'great inflamers' are listed in the next part of the article, published 3 November 1816 (689–690).
39. [Leigh Hunt], 'Disturbances in the Metropolis,' *The Examiner*, 8 December 1816, 769.
40. Barrell, 'Fire,' 293–297.
41. Timothy Clark, *Ecocriticism on the Edge: The Anthropocene as a Threshold Concept* (London: Bloomsbury, 2015), 109, 72.
42. Bruno Latour, 'Agency in the Time of the Anthropocene,' *New Literary History* 45 (2014): 1–18 (8).
43. For 'species thinking' in relation to capital, see Dipesh Chakrabarty, 'The Climate of History: Four Theses,' *Critical Inquiry* 35 (2009): 197–222. For a critique, see Andreas Malm and Alf Hornborg, 'The geology of mankind? A critique of the Anthropocene narrative,' *The Anthropocene Review* 1 (2014): 62–69. For Chakrabarty's response, see 'The Politics of Climate Change is more than the Politics of Capitalism,' *Theory, Culture & Society* 34 (2017): 25–37. For the emissions data, see Timothy Gore, 'Extreme Carbon Inequality', OXFAM Media Briefing, accessed 3 December 2016, http://policy-practice.oxfam.org.uk/publications/extreme-carbon-inequality-why-the-paris-climate-deal-must-put-the-poorest-lowes-582545.
44. Jane Bennett, *Vibrant Matter: A Political Ecology of Things* (Durham and London: Duke University Press, 2010), 37–38.
45. Ian Buchanan, 'What must we do about rubbish?', accessed 3 December 2016, http://drainmag.com/what-must-we-do-about-rubbish/. While Buchanan raises important questions about vital materialism, his critique of Bennett is based on what seems to me a rather selective reading of *Vibrant Matter*.
46. Amitav Ghosh, *The Great Derangement: Climate Change and the Unthinkable* (Chicago and London: The University of Chicago Press, 2016), 63.

Bibliography

Primary

Anon. 'Reform'. *The Examiner*, 19 January 1817.

Anon. 'Political Affairs in November'. *Monthly Magazine*, December 1816.

Anon. 'The Corn Bill'. *The Examiner*, 8 December 1816.

Anon. 'Meeting at Spa-Fields'. *The Examiner*, 17 November 1816.

Anon. 'Art. I. The Speech of Charles C. Western'. *Edinburgh Review*, June 1816.

Anon. 'New Regulations for the Batavian Society, of Arts and Sciences'. *Transactions of the Batavian Society, of Arts and Sciences* 7 (1814): xiv–xvii.

[Assey, Charles.] 'Narrative of the Effects of the Eruption from the Tomboro Mountain, in the Island of Sumbawa, on the 11th and 12th of April 1815'. *Transactions of the Batavian Society, of Arts and Sciences* 8 (1816): 3–25.

———. *Review of the Administration, Value, and State of the Colony of Java*. 1816.

Buffon, George-Louis Leclerc. *Natural History, General and Particular*. Translated by William Smellie. 9 vols. London: A. Strahan and T. Cadell. 3rd edition.

Byron, George Gordon. *The Complete Poetical Works*. Edited by Jerome McGann. 7 vols. Oxford: Clarendon Press, 1980–1993.

———. *Byron's Letters and Journals*. Edited by Leslie A. Marchand. 12 vols. London: John Murray, 1973–1982.

Cobbett, William. 'To the Journeymen and Labourers of England, Wales, Scotland, and Ireland'. *Political Register*, 2 November 1816.

———. 'Corn Crops'. *Political Register*, 31 August 1816.

D. Higgins, *British Romanticism, Climate Change, and the Anthropocene*,
DOI 10.1007/978-3-319-67894-8

Coleridge, Samuel Taylor. *Coleridge's Poetry and Prose*. Edited by Nicholas Halmi, Paul Magnuson, and Raymond Modiano. New York: Norton, 2004.

———. 'Fire, Famine, and Slaughter'. *The Examiner*, 24 November 1816.

Crawfurd, John. *History of the Indian Archipelago*. 3 vols. Edinburgh: Constable, 1820.

Curran, John Philpott. Speech quoted in 'Reform'. *The Examiner*, 19 January 1817.

Defoe, Daniel. *The Storm*. Edited by Richard Hamblyn. London: Penguin, 2005.

De Quincey, Thomas, *The Works of Thomas De Quincey*, ed. by Grevel Lindop *et al.*, 21 vols (London: Pickering and Chatto, 2000–2003).

G. K. 'Want, Famine, and Mortality'. *The Black Dwarf*, 21 May 1817.

Hogg, Thomas Jefferson. *The Life of Percy Bysshe Shelley*. 4 vols. London: Edward Moxon, 1858.

Hunt, Leigh. 'On the Necessity of being Watchful during the Vacation of Parliament'. *The Examiner*, 10 August 2017.

———. 'Disturbances in the Metropolis'. *The Examiner*, 8 December 1816.

———. 'Mr. Hunt of Bristol and the Editor of the Times'. *The Examiner*, 24 November 1816.

———. 'Who are the Inflammatory?' *The Examiner*, 3 November 1816.

———. 'Who are the Inflammatory?' *The Examiner*, 27 October 1816.

———. 'Relief of the National Distresses'. *The Examiner*, 1 September 1816.

———. 'Relief of the National Distresses'. *The Examiner*, 11 August 1816.

Java Half-Yearly Almanac and Directory for 1815. Batavia: A. H. Hubbard, Government Press, 1815. https://sites.google.com/site/sumatraswestkust/java-almanac-1. Accessed on 30 May 2016.

Milton, John. *Paradise Lost*. Edited by Christopher Ricks. London: Penguin, 1968.

Philomath. 'The Works of God Displayed: An Account of a Volcanic Eruption'. *Methodist Magazine* December 1816: 910–911.

Polidori, John William. *Ernestus Berchtold; or the Modern Oedipus. A Tale*. London: Longman, 1819.

Raffles, Sophia, ed. *Memoir of the Life and Public Services of Sir Thomas Stamford Raffles*. London: John Murray, 1830.

Raffles, Thomas Stamford. *The History of Java*. 2 vols. Kuala Lumpur: Oxford University Press, 1965. Reprint of 1817 edition.

———. 'A Discourse Delivered at a Meeting of the Society of Arts and Sciences in Batavia, on the Twenty-fourth day of April 1813, being the Anniversary of the Institution'. *Transactions of the Batavian Society, of Arts and Sciences* 7 (1814): 1–15.

Shakespeare, William. *The Life and Death of King Richard the Second*. In *Complete Works*, edited by Jonathan Bate and Eric Rasmussen. London: Palgrave, 2008.

Shelley, Mary. *Frankenstein*. Edited by D. L. Macdonald and Kathleen Scherf. Peterborough, Ontario: Broadview, 2005. 2nd edition.

———. *The Journals of Mary Shelley, 1814–44*. Edited by Paula R. Feldman and Diana Scott-Kilvert. 2 vols. Oxford: Clarendon Press, 1987.

———. and Percy Bysshe Shelley. *History of a Six Weeks Tour*. London: T. Hookham and C. and J. Ollier, 1817.

Shelley, Percy Bysshe. *The Complete Poetry of Percy Bysshe Shelley*. Edited by Donald Reiman and Neil Fraistat. 3 vols. Baltimore: The Johns Hopkins University Press, 2000–2012.

———. *The Prose Works of Percy Bysshe Shelley*. Edited by E. B. Murray. 1 vol. Oxford: Clarendon Press, 1993.

———. *The Poems of Shelley Volume 1: 1804–17*. Edited by Geoffrey Matthews and Kelvin Everest. London: Routledge, 1989.

———. *The Letters of Percy Bysshe Shelley*. Edited by Frederick L. Jones. 2 vols. Oxford: Clarendon Press, 1964.

———. 'Hymn to Intellectual Beauty'. In 'Original Poetry', *The Examiner*, 19 January 1817.

Southey, Robert. *The Collected Letters of Robert Southey: A Romantic Circles Electronic Edition: Part Five: 1816–1818*. Edited by Tim Fulford, Ian Packer, and Lynda Pratt. http://www.rc.umd.edu/editions/southey_letters/Part_Five/index.html. Accessed on 5 February 2017.

T. 'On the Deficiency of the present Crop, and Economy recommended in its Consumption'. *The Farmer's Magazine*, November 1816.

White, Gilbert. *The Natural History of Selborne*. Edited by Richard Mabey. London: Penguin, 1987.

Secondary

Alaimo, Stacy. 'Elemental Love in the Anthropocene'. In *Elemental Ecocriticism: Thinking with Earth, Air, Water, and Fire*, edited by Jeffrey Jerome Cohen and Lowell Duckert, 298–309. Minneapolis and London: University of Minnesota Press, 2015.

Anderson, Mark D. *Disaster Writing: The Culture Politics of Catastrophe in Latin America*. Charlottesville and London: University of Virginia Press, 2011.

'An Ecomodernist Manifesto', accessed 31 July 2017, http://www.ecomodernism.org/.

Bailes, Melissa. 'The Psychologization of Geological Catastrophe in Mary Shelley's *The Last Man*'. *ELH* 82 (2015): 671–699.

Barad, Karen. *Meeting the Universe Halfway: Quantum Physics and the Entanglement of Matter and Meaning*. Durham and London: Duke University Press, 2007.

———. 'Posthumanist Perfomativity: Towards an Understanding of How Matter Comes to Matter'. *Signs: Journal of Women in Culture and Society* 28 (2003): 801–831.

Barrell, John. '"Fire, Famine, and Slaughter"'. *Huntingdon Library Quarterly* 63 (2000): 276–298.

Bate, Jonathan. *Romantic Ecology: Wordsworth and the Environmental Tradition.* London: Routledge, 1991.

———. *The Song of the Earth.* London: Picador, 2000.

Behringer, Wolfgang. *A Cultural History of Climate.* Cambridge: Polity, 2009.

Bennett, Jane. 'Systems and Things: A Response to Graham Harman and Timothy Morton'. *New Literary History* 43 (2012): 225–233.

———. *Vibrant Matter: A Political Ecology of Things.* Durham and London: Duke University Press, 2010.

Bewell, Alan. 'Jefferson's Thermometer: Colonial Biogeographical Constructions of the Climate of America'. In *Romantic Science: The Literary Forms of Natural History*, edited by Noah Heringman. New York: SUNY Press, 2003, 111–138.

———. *Romanticism and Colonial Disease.* Baltimore: The Johns Hopkins University Press, 1999.

Blanchot, Maurice. *The Writing of the Disaster.* Translated by Ann Smock. Lincoln and London: University of Nebraska Press, 1995.

Bonneuil, Christophe, and Jean-Baptiste Fressoz. *The Shock of the Anthropocene.* Translated by David Fernbach. London: Verso, 2016.

Brassier, Raymond. *Nihil Unbound: Enlightenment and Extinction.* Basingstoke: Palgrave, 2007.

Brewer, William D. *The Shelley-Byron Conversation.* Gainesville: University Press of Florida, 1994.

Bristow, Tom, and Thomas H. Ford, eds. *A Cultural History of Climate Change.* London: Routledge, 2016.

Buchanan, Ian. 'What must we do about rubbish?' http://drainmag.com/what-must-we-do-about-rubbish/. Accessed on 3 December 2016.

Carrigan, Anthony. 'Towards a Postcolonial Disaster Studies'. In *Global Ecologies and the Environmental Humanities: Postcolonial Approaches*, edited by Elizabeth DeLoughrey, Jill Didur, and Anthony Carrigan, 117–139. London: Routledge, 2015.

———. '"Out of this great tragedy will come a world class tourism destination": Disaster, Ecology, and Post-Tsunami Tourism Development in Sri Lanka'. In *Postcolonial Ecologies: Literatures of the Environment*, edited by Elizabeth DeLoughrey and George Handley, 273–290. Oxford: Oxford University Press, 2011.

Carroll, Siobhan. *An Empire of Air and Water: Uncolonizable Space in the British Imagination, 1750–1850.* Philadephia: University of Pennsylvania Press, 2015.

———. 'Crusades Against Frost: *Frankenstein*, Polar Ice, and Climate Change in 1818'. *European Romantic Review* 24 (2013): 211–230.

Chakrabarty, Dipesh. 'Humanities in the Anthropocene: The Crisis of an Enduring Kantian Fable'. *New Literary History* 47 (2016): 377–397.

———. 'The Politics of Climate Change is more than the Politics of Capitalism'. *Theory, Culture & Society* 34 (2017): 25–37.

———. 'Postcolonial Studies and the Challenge of Climate Change'. *New Literary History* 43 (2012): 1–18.

———. 'The Climate of History: Four Theses'. *Critical Inquiry* 35 (2009): 197–222.

Clark, Nigel. *Inhuman Nature: Sociable Life on a Volatile Planet.* London: Sage, 2011.

———. 'Volatile Worlds, Vulnerable Bodies: Confronting Abrupt Climate Change'. *Theory, Culture & Society* 27, 2–3 (2010): 31–53.

Clark, Timothy. *Ecocriticism on the Edge: The Anthropocene as a Threshold Concept.* London: Bloomsbury, 2015.

———. *The Cambridge Introduction to Literature and the Environment.* Cambridge: Cambridge University Press, 2011.

Clubbe, John. 'The Tempest-toss'd Summer of 1816: Mary Shelley's *Frankenstein*'. *Byron Journal* 19 (1991): 26–40.

Cohen, Jeffrey Jerome, and Lowell Duckert, eds. *Elemental Ecocriticism: Thinking with Earth, Air, Water, and Fire.* Minneapolis and London: University of Minnesota Press, 2015.

Colebrook, Claire. *Death of the PostHuman: Essays on Extinction, Vol. 1.* Michigan: Open Humanities Press, 2014.

Craciun, Adriana. 'Writing the Disaster: Franklin and *Frankenstein*'. *Nineteenth-Century Literature* 65 (2011): 433–480.

Cruikshank, Julie. *Do Glaciers Listen? Local Knowledge, Colonial Encounters, and Social Imagination.* Vancouver: UBC Press, 2005.

Davies, Jeremy. *The Birth of the Anthropocene.* Oakland: University of California Press, 2016.

Deleuze, Gilles, and Félix Guattari. *A Thousand Plateaus: Capitalism and Schizophrenia.* Translated by Brian Massumi. London: Bloomsbury, 2016.

DeLoughrey, Elizabeth, Jill Didur, and Anthony Carrigan, eds. *Global Ecologies and the Environmental Humanities: Postcolonial Approaches.* London: Routledge, 2015.

DeLoughrey, Elizabeth, and George Handley, eds. *Postcolonial Ecologies: Literatures of the Environment.* Oxford: Oxford University Press, 2011.

Dingley, R. J. '"I had a Dream…..": Byron's "Darkness"'. *Byron Journal* 9 (1981): 20–33.

Duckert, Lowell. 'Glacier'. *postmedieval: a journal of medieval cultural studies* 4 (2013): 68–79.

Duffy, Cian. *Shelley and the Revolutionary Sublime.* Cambridge: Cambridge University Press, 2005.

Economides, Louise. '"Mont Blanc" and the Sublimity of Materiality'. *Cultural Critique* 61 (2005): 87–114.

Ellermann, Greg. 'Speculative Romanticism'. *SubStance* 44.1 (2015): 154–174.

Fagan, Brian. *The Little Ice Age: How Climate Made History 1300–1850.* New York: Basic, 2000.

Favret, Mary A. *War at a Distance: Romanticism and the Making of Modern Wartime.* Princeton and Oxford: Princeton University Press, 2010.

Ferguson, Frances. 'Shelley's *Mont Blanc*: What the Mountain Said'. In *Romanticism and Language*, edited by Arden Reed, 202–214. London: Methuen, 1984.

Fulford, Tim. *Romantic Poetry and Literary Coteries: The Dialect of the Tribe.* New York: Palgrave, 2015.

Ghosh, Amitav. *The Great Derangement: Climate Change and the Unthinkable.* Chicago and London: The University of Chicago Press, 2016.

Gidal, Eric. *Ossianic Unconformities: Bardic Poetry in the Romantic Age.* Charlottesville and London: University of Virginia Press, 2015.

———. '"O Happy Earth! Reality of Heaven!": Melancholy and Utopia in Romantic Climatology'. *Journal for Early Modern Cultural Studies* 8 (2008): 74–101.

Glendinning, Victoria. *Raffles and the Golden Opportunity.* London: Profile, 2012.

Gore, Timothy. 'Extreme Carbon Inequality'. OXFAM Media Briefing. http://policy-practice.oxfam.org.uk/publications/extreme-carbon-inequality-why-the-paris-climate-deal-must-put-the-poorest-lowes-582545. Accessed on 3 December 2016.

Golinski, Jan. *Weather and the Climate of Enlightenment.* Chicago and London: The University of Chicago Press, 2007.

Gottlieb, Evan. *Romantic Realities: Speculative Realism and British Romanticism.* Edinburgh University Press, 2016.

Griffin, Andrew. 'Fire and Ice in Frankenstein'. In *The Endurance of* Frankenstein*: Essays on Mary Shelley's Novel*, edited by George Levine and U. C. Knoepflmacher. Berkeley: University of California Press, 1979, 49–73.

Hamblyn, Richard. *Terra: Tales of the Earth.* London: Picador, 2009.

Hamilton, Clive. 'The Anthropocene as rupture'. *The Anthropocene Review* 3 (2016): 93–106.

Hamilton, Clive and Jacques Grinewald. 'Was the Anthropocene anticipated?'. *The Anthropocene Review* 2 (2015): 59–72.

Harman, Graham. *Guerrilla Metaphysics: Phenomenology and the Carpentry of Things.* Chicago: Open Court, 2005.

———. 'The Well-Wrought Broken Hammer: Object-Orientated Literary Criticism'. *New Literary History* 43 (2012): 183–203.

Harris, Alexandra. *Weatherland: Writers & Artists Under English Skies.* London: Thames & Hudson, 2015.

Heringman, Noah. 'Deep Time at the Dawn of the Anthropocene'. *Representations* 129 (2015): 56–85.

———. *Romantic Rocks, Aesthetic Geology.* Ithaca and London: Cornell University Press, 2004.

———. ed. *Romantic Science: The Literary Forms of Natural History.* New York: SUNY Press, 2003.

Hornborg, Alf. 'Artifacts have consequences, not agency: Toward a critical theory of global environmental history'. *European Journal of Social Theory* 20 (2016): 1–16.

Hubbell, Andrew. 'Our Mix'd Essence': Manfred's Ecological Turn'. *Byron Journal* 42 (2014): 5–20.

Hulme, Mike. *Weathered: Cultures of Climate.* London: Sage, 2017.

———. *Why We Disagree About Climate Change: Understanding Controversy, Inaction, and Opportunity.* Cambridge: Cambridge University Press, 2009.

Ingrams, Richard. *The Life and Adventures of William Cobbett.* London: Harper Perennial, 2006.

Iovino, Serenella, and Serpil Oppermann. 'Introduction'. In *Material Ecocriticism*, edited by Serenella Iovino and Serpil Oppermann, 1–17. Bloomington and Indianapolis: Indiana University Press, 2014.

———. eds. *Material Ecocriticism.* Bloomington and Indianapolis: Indiana University Press, 2014.

Jasanoff, Sheila. 'A New Climate for Society'. *Theory, Culture & Society* 27 (2010): 233–253.

Johns-Putra, Adeline. 'Historicizing the Networks of Ecology and Culture: Eleanor Anne Porden and Nineteenth-Century Climate Change'. *Interdisciplinary Studies in Literature and Environment* 22 (2015): 27–46.

Jones, Christine Kenyon. '"When this world shall be *former*": Catastrophism as imaginative theory for the younger Romantics'. *Romanticism on the Net* 24 (2001). 16 paragraphs. http://www.erudit.org/revue/ron/2001/v/n24/006000ar.html. Accessed on 22 March 2016.

Jong Boers, Bernice de. 'Mount Tambora in 1815: A Volcanic Eruption in Indonesia and its Aftermath'. *Indonesia* 60 (October 1995): 37–59.

Kandlbauer, J., P. O. Hopcroft, P. J. Valdes, and R. S. J. Sparks. 'Climate and carbon cycle responses to the 1815 Tambora volcanic eruption'. *Journal of Geophysical Research: Atmospheres* 118 (2013): 12497–12507.

Kapstein, I. J. 'The Meaning of Shelley's "Mont Blanc"'. *PMLA* 62 (1947): 1046–1060.

Keach, William. *Shelley's Style*. Methuen: New York and London, 1984.

Khalip, Jacques, and David Collings. 'Introduction: The Present Time of "Live Ashes"'. In *Romanticism and Disaster*, edited by Jacques Khalip and David Collings. *Romantic Circles*. January 2012. http://www.rc.umd.edu/praxis/disaster/HTML/praxis.2012.khalip.html. Accessed on 22 September 2016.

Klingaman, William K., and Nicholas P. Klingaman. *The Year Without A Summer: 1816 and the Volcano that Darkened the Sky and Changed History*. London: St. Martin's Press, 2013.

Kroeber, Karl. *Ecological Literary Criticism: Romantic Imagining and the Biology of Mind*. New York: Columbia University Press, 1994.

Lapp, Robert. *Contest for Cultural Authority: Hazlitt, Coleridge, and the Distresses of the Regency*. Detroit: Wayne State University Press, 1999.

Latour, Bruno. 'Agency at the Time of the Anthropocene'. *New Literary History* 45 (2014): 1–18.

———. 'Love Your Monsters: Why We Must Care For Our Technologies As We Do Our Children'. *Breakthrough Journal* 2 (2012). http://thebreakthrough.org/index.php/journal/past-issues/issue-2/love-your-monsters. Accessed on 8 May 2016.

Lau, Beth, ed. *Fellow Romantics: Male and Female British Writers, 1790–1835*. Farnham: Ashgate, 2009.

Leask, Nigel. 'Mont Blanc's Mysterious Voices: Shelley and Huttonian Earth Science'. In *The Third Culture: Literature and Science*, edited by Elinor S. Shaffer. Berlin and New York: Walter de Gruyter, 1998, 182–203.

Leighton, Angela. *Shelley and the Sublime: An Interpretation of the Major Poems*. Cambridge: Cambridge University Press, 1984.

Levine, George, and U. C. Knoepflmacher, eds. *The Endurance of* Frankenstein*: Essays on Mary Shelley's Novel*. Berkeley: University of California Press, 1979.

Linthicum, Kent. '"Bread or Blood": Climate Insecurity in East Anglia in 1816'. http://www.historicalclimatology.com/blog/bread-or-blood-climate-insecurity-in-east-anglia-in-1816. Accessed on 2 June 2017.

Malm, Andreas and Alf Hornborg. 'The geology of mankind? A critique of the Anthropocene narrative'. *The Anthropocene Review* 1 (2014): 62–69.

McKusick, James. *Green Writing: Romanticism and Ecology*. Basingstoke: Palgrave, 2000.

Mee, John. *Conversable Worlds: Literature, Contention and Community, 1762–1830*. Oxford: Oxford University Press, 2011.

Meillassoux, Quentin. *After Finitude: An Essay on the Necessity of Contingency*. Translated by Ray Brassier. London: Bloomsbury, 2008.

Mikhail, Alan. 'Enlightenment Anthropocene'. *Eighteenth-Century Studies* 49 (2016): 211–231.

Monteith, Sharon, ed. *Hurricane Katrina: Five Years After*. Special issue of the *Journal of American Studies* 44 (2010).

Morton, Timothy. 'Byron's Manfred and Ecocriticism'. In *Palgrave Advances in Byron Studies*, edited by Jane Stabler. London: Palgrave, 2007, 155–170.

———. *Hyperobjects: Philosophy and Ecology After the End of the World*. Minneapolis: University of Minnesota Press, 2013.

———. 'Romantic Disaster Ecology: Blake, Shelley, Wordsworth'. In *Romanticism and Disaster*, edited by Jacques Khalip and David Collings. *Romantic Circles*. January 2012. http://www.rc.umd.edu/praxis/disaster/HTML/praxis.2012.morton.html. Accessed on 22 September 2016.

———. *Shelley and the Revolution in Taste: The Body and the Natural World*. Cambridge: Cambridge University Press, 1994.

———. *The Ecological Thought*. Cambridge, MA: Harvard University Press, 2008.

Murphy, Olivia. 'Apocalypse Not Quite: Romanticism and the Post-human World'. In *Romanticism and Sustainability: Endurance and the Natural World, 1780–1830*, edited by Ben P. Robertson. Lexington Books: Lanham MD, 2015, 245–259.

Nanni, Giordano. *The Colonisation of Time: Ritual, Routine, and Resistance in the British Empire*. Manchester and New York: Manchester University Press, 2012.

Nattrass, Leonora. *William Cobbett: The Politics of Style*. Cambridge: Cambridge University Press, 1995.

Neiman, Susan. *Evil in Modern Thought: An Alternative History of Philosophy*. Princeton: Princeton University Press, 2002.

Nixon, Rob. *Slow Violence and the Environmentalism of the Poor*. Cambridge, MA: Harvard University Press, 2011.

Oliver, Susan. 'Ice Thoughts 2'. 9 November 2016. https://susanoliverweb.com/2016/11/09/ice-thoughts-2/. Accessed on 20 January 2017.

Oppenheimer, Clive. *Eruptions that Shook the World*. Cambridge: Cambridge University Press, 2011.

———. 'Climatic, environmental and human consequences of the largest known historical eruption: Tambora volcano (Indonesia) 1815'. *Progress in Physical Geography* 27 (2003): 230–259.

Parker, Geoffrey. *Global Crisis: Climate Change and Catastrophe in the Seventeenth Century*. New Haven and London: Yale University Press, 2013.

Post, John D. *The Last Great Subsistence Crisis in the Western World*. Baltimore and London: The Johns Hopkins University Press, 1977.

Pulos, C. E. *The Deep Truth: A Study of Shelley's Scepticism*. Lincoln: University of Nebraska Press, 1962.

Redford, Catherine. '"No love was left": The Failure of Christianity in Byron's "Darkness"'. *Byron Journal* 43 (2015): 131–140.

Reed, Arden, ed. *Romanticism and Language*. London: Methuen, 1984.

Richard, Jessica. '"A Paradise of My Own Creation": *Frankenstein* and the Improbable Romance of Polar Exploration', *Nineteenth-Century Contexts* 25 (2003): 295–314.

Rigby, Kate. *Dancing with Disaster: Environmental Histories, Narratives, and Ethics for Perilous Times.* Charlottesville and London: University of Virginia Press, 2015.

Robinson, Charles E. 'Byron and Mary Shelley and *Frankenstein*'. University of Nottingham: Byron Centre, 2000. http://byron.nottingham.ac.uk/resources/digital/foundation%20lectures/Robinson.pdf. Accessed on 26 March 2016.

———. *Byron and Shelley: The Snake and the Eagle Wreathed in Flight.* Baltimore: The Johns Hopkins University Press, 1976.

Robertson, Ben P. ed. *Romanticism and Sustainability: Endurance and the Natural World, 1780–1830.* Lexington Books: Lanham MD, 2015.

Roe, Nicholas. *Fiery Heart: The First Life of Leigh Hunt.* London: Pimlico, 2005.

Rudwick, Martin. *Bursting the Limits of Time: The Reconstruction of Geohistory in the Age of Revolution.* Chicago: University of Chicago Press, 2005.

———. ed. *Georges Cuvier, Fossil Bones, and Geological Catastrophes: New Translations & Interpretations of Primary Texts.* Chicago and London: The University of Chicago Press, 1997.

Schuster, Joshua. 'How to Write the Disaster'. *Minnesota Review* 83 (2014): 163–171.

Scranton, Roy. *Learning to Die in the Anthropocene: Reflections on the End of a Civilisation.* San Francisco: City Lights Books, 2015.

Shaffer, Elinor S. ed. *The Third Culture: Literature and Science.* Berlin and New York: Walter de Gruyter, 1998.

Shaviro, Steven. *The Universe of Things: On Speculative Realism.* Minneapolis: The University of Minnesota Press, 2014.

Smith, Bruce D. and Melinda A. Zeder. 'The onset of the Anthropocene'. *Anthropocene* 4 (2013): 8–13.

Smith, Victoria C. 'Volcanic markers for dating the onset of the Anthropocene'. In *A Stratigraphical Basis for the Anthropocene*, edited by C. N. Waters, J. A. Zalasiewicz, M. Williams, M. A. Ellis, and A. M. Snelling, 283–299. London: Geological Society Special Publications, 2013.

Somervell, Tess. 'Mediating Vision: Wordsworth's Allusions to Thomson's *Seasons* in *The Prelude*'. *Romanticism* 22 (2016): 48–60.

Stabler, Jane, ed. *Palgrave Advances in Byron Studies.* London: Palgrave, 2007.

Stafford, Fiona. *The Last of the Race: The Growth of a Myth from Milton to Darwin.* Oxford: Oxford University Press, 1994.

Szerszynski, Bronislaw. 'Reading and Writing the Weather: Climate Technics and the Moment of Responsibility'. *Theory, Culture and Society* 27 (2010): 9–30.

Thacker, Eugene. *In The Dust of this Planet: Horror of Philosophy, Vol. 1.* Winchester: Zero Books, 2010.

Vail, Jeffrey. '"The Bright Sun Was Extinguish'd": The Bologna Prophecy and Byron's "Darkness"'. *Wordsworth Circle* 28 (1997): 183–192.

Van Niel, Eloise Smith. *The Batavian Society and Scholarship in the Dutch East Indies, 1778–1850.* MA Thesis, University of Hawaii, May 1980.

Veale, Lucy and Georgina H Endfield. 'Situating 1816, the "year without summer", in the UK'. *The Geographical Journal*: 182 (2016): 318–330.

Washington, Chris. 'Romanticism and Speculative Realism'. *Literature Compass* 12.9 (2015): 448–460.

Wasserman, Earl R. *Shelley: A Critical Reading*. Baltimore and London: The Johns Hopkins University Press, 1971.

Waters, C. N., J. A. Zalasiewicz, M. Williams, M. A. Ellis, and A. M. Snelling, eds. *A Stratigraphical Basis for the Anthropocene.* London: Geological Society Special Publications, 2013.

White, R. J. *Waterloo to Peterloo.* Harmondsworth: Penguin, 1957.

Wilson, Eric G. *The Spiritual History of Ice.* Basingstoke: Palgrave, 2003.

Witze, Alexandra, and Jeff Kanipe. *Island on Fire: The Extraordinary Story of Laki, the Volcano that Turned Eighteenth-Century Europe Dark.* Place: Profile, 2014.

Wood, Gillen D'Arcy, ed. *Eco-Historicism.* Special issue of *The Journal for Early Modern Cultural Studies* 8 (2008).

———. *Tambora: The Eruption that Changed the World.* Princeton: Princeton University Press, 2014.

———. 'The Volcano Lover: Climate, Colonialism, and the Slave Trade in Raffles's *History of Java* (1817)'. *Journal for Early Modern Cultural Studies* 8 (2008): 33–53.

Zalasiewicz, Jan, Mark Williams, *et al.* 'Are we now living in the Anthropocene?'. *GSA Today* 18 (2008): 4–8.

Zilberstein, Anya. *A Temperate Climate: Making Climate Change in Early America.* Oxford: Oxford University Press, 2016.

Index

D. Higgins, *British Romanticism, Climate Change, and the Anthropocene*,
DOI 10.1007/978-3-319-67894-8

The manufacturer's authorised representative in the EU is Springer Nature Customer Service Centre GmbH, Europaplatz 3, 69115 Heidelberg, Germany. If you have any concerns regarding our products, please contact ProductSafety@springernature.com

Printed and bound by CPI Group (UK) Ltd, Croydon, CR0 4YY
25/09/2026
02222890-0015